Contents

PART ONE v

Dedications vii
Preface 1
1 Of Preparation for the Unimaginable 3
2 Of Birth, Family,
   a Little Boy & a Young Man 5
3 Of Writing & Creativity 9
4 Of a Little Brother & Best Buddy 13
5 Of First Friendships & Growing Up 15
6 Of Lasting Friendships 19
7 Of High School Rugby ...
   or Of Giving 100% 23
8 Of Tattoos & Moustaches 25
9 Of Dating & Hurricane Katrina 29
10 Of Relationships That Endured
   Time & Even Death 33
11 Of James' Resolve to Stay In
   Touch & Tie Up Loose Ends 37
12 Of World Teach
   & Marshall Islands Life 41
13 Of First Responders in the RMI—
   The Coast Guard & the Pacific Effort 51
14 Of First Responders
   in Louisiana 55
15 Of Learning About Things We
   Never Wanted to Know 59
16 Of Memorials 63
17 Of the League 71
18 Of Visitors from World Teach
   Who Traveled to See Us in Louisiana 73
19 Of James' Namesakes 79
20 Of the Fruits of James' Life ...
   In the Marshall Islands (RMI) 83
21 Of the Fruits of James' Life ...
   In Mozambique 89
22 Of the Fruits of James' Life ...
   In Louisiana 93
23 Of Life Going On & Arrival
   at the 10-Year Mark 97
24 Of Narratives of Faith 109
25 Of Joy, Thanksgiving, & Peace 117

PART TWO - Thank Yous from
James' Family 119

PART THREE - James' Letters 125

PART ONE

Dedications

In memory of our son, James Clifton deBrueys

In memory of Kevin Heffron, James' uncle & my brother

In memory of the children of friends in my grief group, *Cracked Pots … Blessed & Broken*

In memory of James' grandparents, Mary T White & Norman Clifton Heffron and Mary Louise Henderson & Warren Claude deBrueys

In memory of Katherine Cardinale, the sister of our son-in-law, John

In honor of Jim's and my families

In memory of Kaylee Crousillac, who dated James

In memory of Jonathan Caluda, one of James' high school buddies

In honor of all volunteers for World Teach, especially those who served with James

In honor of all of James' precious friends from childhood through adulthood

In honor of Jim, my husband, and of James' siblings (Michelle, André, Steven & Simone) and brothers-in-law (Patrick, John & Jon)

In honor of James' niece & nephew he knew-Madison & Blake-and in honor of James' nieces & nephew he never met-Samantha James, Nizza, Luke & Emma

In gratitude to everyone who suggested and encouraged this project over the last ten years, particularly to Glenn & Mary Lu Penton, who cheered me on to write and publish James' "letters from paradise", & to Simone for her assistance with photos

In honor of ALL who supported and prayed for us all these years … and who continue to do so

Preface

One of our two sons, James Clifton deBrueys, died on November 25, 2010, almost exactly one month before his twenty-third birthday. Jim and I raised five children, and James was our third. As are each of our children, he was amazing and unique. James was a gift to many people, and the tenth anniversary of our losing him at sea is my opportunity to share him here with you.

Throughout this book, as his mother, I am writing from my perspective in the first person.

Part 1 of the book is anecdotal, filled with pictures. It shares James' life before he went to the Marshalls and when he lived in the RMI (Republic of the Marshall Islands), and then shares a glimpse into our lives over the ten years since he was lost at sea. I force myself to describe what happened and how we have felt and dealt as openly as I can. As a parent, I still find it difficult to use words like "our son's death" and "when our son drowned." For me to say and write these loaded words is both therapeutic and sad.

Part 2 of the book thanks many people. You will see a list of hundreds of people's names alphabetically by first name. These are some of the people mentioned within the book, but mostly they are names of people who have consistently called, prayed, sent cards and notes, and displayed other kindnesses as they walked this journey with our family over these first ten years. If we got a call or card or anything from you at any time, I hope you find your name here. My greatest anxiety in completing this list and pressing the submit button on my computer is that I left someone - even just one person - out. *If I did, I am truly sorry and hope I will be forgiven that oversight.*

Part 3 is solely James' letters in his words as he wrote to family and friends from the Marshall Islands during his few months there. He loved to write, and you will see that he was prolific.

JAMBOS is the Marshallese word for "walks." James enjoyed *jambos* with his island family and students in the Marshall Islands. You will see in his letters that exactly one week before his death, he suggested that one day he might collate all of his letters and might even call them *Jambos with James*. (See James' letter dated 11/18/10.)

Finally, Through this book, my desires are simple and three-fold: 1) to remember James with family and friends; 2) to let you come to feel you know James if you didn't before; and 3) to help family, friends, and each reader see the beauty (and even joy) that can result from such tragedy through the eyes of faith.

Enjoy!

Mary T

1

Of Preparation for the Unimaginable

> Your children are not your children.
> They are the sons and daughters of Life's longing for itself.
> They come through you but not from you,
> And though they are with you yet they belong not to you.
>
> You may give them your love but not your thoughts,
> For they have their own thoughts.
> You may house their bodies but not their souls,
> For their souls dwell in the house of tomorrow,
> which you cannot visit, not even in your dreams.
> You may strive to be like them,
> but seek not to make them like you.
> For life goes not backward nor tarries with yesterday.
>
> You are the bows from which your children
> as living arrows are sent forth.
> The Archer sees the mark upon the path of the infinite,
> and He bends you with His might
> that His arrows may go swift and far.
> Let your bending in the Archer's hand be for gladness;
> For even as He loves the arrow that flies,
> so He loves also the bow that is stable.
>
> - Kahlil Gibran, *The Prophet* (**Chapter:** "On Children")

Memorized years before, these words flooded back to me at the time of our surreal loss in 2010. Sr. Dianne Fanguy, CSJ taught me at St. Joseph's Academy (my high school in Baton Rouge, Louisiana) in the 1970s. For me, one of Sr. Dianne's most memorable lessons included the exploration of this beautiful writing by Gibran. I found the words beautiful and flowing, and the thoughts and precious words of it were appreciated by me even then. The passage's beauty, however, was deeply, fully recognized only after James died.

Some time had passed after we lost our son when Jim and I ran into Sr. Dianne. She cried with us when I reminded her of her teaching. We reflected on the gift of learning something beautiful—if not so pertinent at one point in life—only to have it come back with unfathomable significance and meaning at another.

2

Of Birth, Family, a Little Boy & a Young Man

James entered this world on December 23, 1987 with his umbilical cord wrapped around his neck. Since he was born by emergency Caesarian section, he and I spent his first Christmas in the hospital. He was handed to me in the hospital swaddled inside a little stocking! On Christmas morning, James' dad came to the hospital with our two older daughters, Michelle and André. I remember that they brought me a cappuccino from home. Earlier that morning while holding James, I had listened to our daughters opening Christmas gifts over the phone from my hospital bed. All felt right with the world.

In part, the world feeling "right" could be attributed to my surreal experience the night before, on Christmas Eve, as I lay sleepless in my hospital room. I was sitting up in my bed (the bed closer to the window in the room that was designed for two patients, though the first bed was empty) sipping a cup of hot tea and reading from a small prayer book that offered daily entries. The quote across the top of that day's page was, "Let go and let God." This hit home for me since Jim, my husband, was in danger of losing his position at the hospital where he worked. It was a sort of last-in-first-out situation for him. Now, having just had surgery, I would be later than anticipated returning fully to work myself while Jim might not have a job to which to return. I was anxious about going home with our third child and with a complete lack of control over work for both Jim and for me. As I was reading, a light appeared in the far corner of my room near the window. I looked up, saw Jesus, looked away, and looked back again to an empty space. The apparition was gone, but it was real. I know this because the apparition was replaced by a feeling of extreme peace and a complete lack of anxiety. I recall returning home rid of worry and with no fear or anxiety (about the apparition or any circumstances surrounding us regarding loss of work and possibly of wages). That amazing Presence stayed with me overwhelmingly for several weeks during which our "storms" receded and life went forward.

Over the coming years, try as I might, I have not been able to conjure that incredible feeling of peace. I remember wondering then and over the years why I had been given such a gift after the birth of James and not after the births of our four other children. That was likely the first virtually palpable sign I had that God was truly with me. But, like Gibran's writing, I could not fully appreciate the significance of that awesome occurrence in the hospital just yet. Looking back, both *The Prophet* and my hospital miracle seem to have been God preparing me for what was to come. Some call these special events "God-winks."

Life went on without full appreciation of these God-winks until almost twenty-three years later. To this day, I think about my awareness of being so utterly anxiety-less back in 1987. I want so much to feel that again. I have come to believe it was a gift, preparing me for the loss of James in the future and helping me to believe—to know—that the peace I felt will once again be possible. It develops, it evolves, it increases as we move on in this life without one of our children who is so precious to us. I believe we will all be together completely and forever with the communion of saints in the fullness of time.

The friends James made when he was a little boy continue to be our friends to this day. What gifts, for example, emerged from the first three outside caregivers James knew. Frances Linker was the first lady who helped care for him outside of our home when he was a baby and toddler. She cared for only a few children at a time in her home, and she lived near a school where I saw students for parts of my work days. When James was with her, his little friends were Chaz, Jason, Elizabeth, Brittany, and Alex Grush. Traci Roussel Gerald was our go-to babysitter in our home at that time, and James later attended Seven Oaks Academy pre-kindergarten. It was at Seven Oaks that he met two dear, life-long friends, Rhett Ward and Carmen Linares. Now, as I type this entry, I am overcome with the precious relationships Jim and I have with Frances, Traci, Carmen, Rhett, and their families.

To young people who I hope are reading and enjoying these memories, please be assured that connections you maintain with friends' parents as they (and you) get older are treasures.

Jim and I loved having our children and watching them grow up. They were the focus of our lives throughout the many years of raising them, and they remain so now, along with our grandchildren. As parents, certain memories stand out and are replayed by us and our children when we gather and talk about special times now that we are all older. One of our favorite memories was of Jim and me telling the kids one Christmas that our family gift to them was going to be a trip to Disney World. Jim and I made a scroll that had to be unrolled, and on Christmas morning, Michelle, as our oldest, read the message announcing the great surprise. Their reactions revealed perfectly their personalities. Michelle quickly said she was overjoyed to miss some days of school. I remember that André had a bad cold that morning, but she let us know with her deteriorating voice that she was excited about the trip. James, Steven, and Simone were five, three, and two years old at the time. Steven jumped around saying, "Ride 'em, cowboy!" in his excitement (obviously a cartoon fan at his young age). Simone, still so young, smiled and looked a bit confused as to what was going on, though she seemed to understand that it was a good thing. And James' immortal line was, "But I don't want to miss school!" He was a conscientious little guy even then. Over the years when we remembered this special time as a family or watch the video of our children's reactions, we recall our other four ribbed James with this now classic line, "But I don't want to miss school!"

James valued academics, reading, and languages. He quickly displayed a talent for writing and an unfolding and ever-evolving imagination.

3

Of Writing & Creativity

James' love of writing and his creativity are obvious in his *JAMBOS* letters printed in this book, but this propensity was equally obvious in school, in notes, through letters he wrote to others, through his journals that introduced imaginary countries and characters, and in movie scripts written with friends as a little boy.

Growing up, James had two best friends on our Kenner street, Brett Bordelon and Chris Vizzini. I remember that their parents and we once expressed—since the three seemed inseparable—that they would probably each be in each other's weddings one day. With the assistance of Steven, James' little brother, these little boys created their own private venture, a film company called Blurry Monkey Productions. With Steven (and sometimes another neighborhood friend or two allowed to join the project of the day), they canvased the neighborhood for settings for their movies. They filmed in each other's homes and attics, filling their movies with their antics. They traveled by foot, bike, carts, and whatever other mode of transportation they could conjure. We are grateful to have some of their Blurry Monkey Production movies on tape to this day.

At his high school graduation ceremony, James earned an award for outstanding English performance. He had a voracious appetite for cultures and languages, studying Spanish and Latin in school and Japanese with a tutor named Johji. He also studied Arabic on his own and began mastering Marshallese before his stint in the Marshall Islands. But James had spoken so incoherently when he was first learning to speak that we called his speech *Jamese* (much as his brother had; hence *Stevenese*). For being two such "language-brained" boys, it intrigued me that their English was unrecognizable for a period of time when they were small. Then there was the conlang Hypervarian language.

Are you curious about Hypervaria? Noteworthy, throughout junior high and high school, was James' creation of an evolving, imaginary country called Hypervaria with a corresponding oral and written language (Hypervarian). James even fashioned a national anthem and coat of arms for his country. In high school, the story was that James convinced one of his teachers, through his exhaustive, detailed, and probably relentless sharing of information about Hypervaria, that this country and language actually existed.

James' creation of a cartoon-like character, Snee, was also a hit. After his death, his girlfriend, Kaylee, had a likeness of Snee tattooed on her wrist. The legend of Snee continues to this day.

James' creativity and bent for being on-show was demonstrated through his involvement at KLSU, the on-campus radio station at Louisiana State University, during his college years. Other special individuals mentioned in this memoir, including Emily Eck, Erin Elizabeth Mikulak, and Grace Doll, worked there as well. James had a perfectly suited voice as a DJ, which he could modify to sound quite professional. He made a number of commercials for local businesses through the station. A taped compilation of most, if not all, of those commercials that we found after his death is now a priceless gift to us.

After he died, our dear next-door neighbors from Kenner, Lelia and Joe Prange, brought us a note they had saved that had been written by James when he'd graduated from high school. Lelia said she had not wanted to discard it because it was funny … "So James." His sense of humor (and his handwriting) never changed much.

To me, James was an intriguing combination of introvert and extrovert. I remember that he went through phases of not wanting to have his picture taken and of not wanting to smile. And yet, he was funny, entertaining, and seemed eager to perform.

Finally, James' ingenuity included a musical bent. He dabbled a bit in harmonica, flute, and ukulele, took lessons to play the sax for two years when he was younger (which he didn't like and that his dad required him to complete), and was in a band with several friends on campus. This group played for events such as fundraisers and for friends. Two of his closest friends in this enterprise were Craig Clement and Adam Delaune. James didn't play an instrument, but he sang with enthusiasm. Steven, his younger brother, later described James' voice as "majestic and regal!" With the passage of time and the intense loyalty of a brother, I would expect nothing less of a description.

4

Of a Little Brother & Best Buddy

Steven, James' younger brother, doubled as James' best friend from the time they were small. They were born only eighteen months apart. When Steven came home from the hospital, I have notes in James' baby book that say how excited James was and that he kept kissing Steven and anyone else in the room. I wrote that it was "so cute" that James chuckled and laughed "a lot and hard" with everyone in the room, especially Steven.

One of our most precious memories of James and Steven as little boys was the summer that their two baseball teams were in our city's playoff against each other. What a family conflict! Jim and I sat on alternate sides of the field per inning and hollered for whichever team was at bat. For this auspicious occasion, our conflict's resolution was ideal if we do say so ourselves. Steven's team won the championship, and James balanced that win by catching a fly ball toward the end of the game to get Steven out. We concluded this was a win-win from a family perspective.

James and Steven, with Simone, their younger sister, competed in chess together locally, at the state level, and nationally through their school under the direction of St. Elizabeth Ann Seton's amazing head coach, Dave Pierson. They competed on teams comprised of school friends and became quite learned in chess. A number of awards followed for all three of them. Our travels with our three youngest took us to a number of states for tournaments. When they were hanging out at our home on weekends or after school, our kids and their neighborhood friends frequently sat down to games of chess in our den. I admired them so, in particular because I (unlike their dad) have not learned the game to this day.

5

Of First Friendships & Growing Up

James and his best buddies on our street (Monte Carlo Drive in Kenner, Louisiana) were close—both in neighborhood proximity and in kindred spirit. This lasted throughout elementary school, high school, and beyond. We moved to that street when James was three from nearby Indiana Avenue in the same town. Together these three little boys were the "three amigos." Chris Vizzini and Brett Bordelon explored our neighborhood with James, making videos, competing in and traveling with their school chess teams, playing ball, creating adventures both real and imaginary, sharing strength in academics, going to NASA's Space Camp and Aviation Challenge, participating in Scouts, discovering video games, beginning to date, and working their first jobs.

If you didn't know James before, through reading *JAMBOS* you will come to have no doubt that James valued friends tremendously. Just as he made new friends in the RMI, he left many behind in the States. There are a few fast friends who stand out from pre-kindergarten, elementary years, and high school. Chris, Brett, Rhett, Steven, James, Craig, Caz, Jason, and others … these boys developed strong bonds, most of which continued throughout the years and with our family to the present day.

High school provided outlets for James to explore his inclination for helping others. He is described and remembered by many as having a tender heart. James participated in the Big Buddy program through his school, played rugby, and studied clandestinely. Friends often commented that James didn't seem to have to study much. This was not the case. He was an exemplary time manager, so he had time for what he wanted to do outside of school work. Even his siblings used to say he didn't study much; actually, he studied during times when others didn't see him or realize that's what he was up to. He bussed tables at a local restaurant and was good at saving his money. Jumping ahead in time to college graduation, I remember James telling me he was glad we didn't give him unlimited credit card usage and that we required him to buy his own car and pay for gas and related expenses. He said that, now that he was graduating, he was proud of himself for what he accomplished. With our family of five children, I was grateful for that feedback, which countered my self-doubt and wishing that maybe we could have/would have given him and our other children more.

Among my sweetest memories of James are times he walked into my office to tell me something or to ask something of seemingly little importance, but then would sit down as longer, deeper conversations evolved. In this very room where I am now typing these memories, he sat at a desk near the window and opened up about friends, school, dating, and antics when he was in the mood to do so. He wasn't always inclined to do this. I learned that, when he was, I wanted to stop what I was doing to listen and engage in this special time of bonding with him. His mind held deep thoughts, especially related to family, friends, feelings, loyalty, betrayal, joy, hurt, various faiths, his faith, and his future.

I feel like it is fair to say that James was complicated, representing a gumbo of attributes. He was both self-assured and insecure. I remember that he would not get on the Space Mountain ride even as an older child. He sat down below on a bench with me (who also didn't want to get on that ride) and waited for the others to come off. I remember that he felt disappointed in himself, but was just plain afraid. He redeemed himself in college when he got on that ride and on the Tower of Terror at Disney World with his girlfriend at the time, Emily. I am willing to bet he was nervous then, too.

6

Of Lasting Friendships

As James and his closest friends became young men, they stuck together. Many of them attended LSU.
One of James' highlights during college was his trip to Spain and Portugal through the LSU Study Abroad Program with fellow classmates and Professor Castro. Later, his sister and her husband, André and John, went to Spain to teach, as did a friend of James', Cody, a short time later. André and Cody each took a photo with a particular bronze statue in the center of Madrid. All were taken on different days in different years, the last two intentionally taken to imitate the photos we have from that specific Madrid location where James stood.

In the summer of 2011, a few months after James was lost at sea, a friend named Ian Brown came to see Jim and me to deliver a special project he'd designed, a book called *El Chico Con La Barba Conquers The Iberian Peninsula*. On the inside back cover of the book, Ian had written: "You, as his most loved ones, know fully the nature of James' impact on others. I know you probably hear praise of James often, but I don't think you can ever hear it enough. Some of the greatest memories of my life include and are thanks to James. To forget James would be to forget a part of myself. I will always remember James and that smile too big and brilliant to be contained even by the world's 4th best beard. Sincerely, Ian Brown." (For an explanation of the beard comment, see chapter titled OF RELATIONSHIPS THAT ENDURED TIME & EVEN DEATH.) Ian told us that the landlady where they stayed in Granada had trouble remembering James' full name, so she referred to him as *"el chico con la barba* (the boy with the beard)."

Ian told us that James and his friends studied and explored together during this memorable trip. Others on the trip that summer included Emily Perkins (who delivered the book to us later with Ian and who, with Ian, on that day that they visited showed us their tattoos in memory of James), along with Alexandra Goodwin, Andrew Herpich, Kristie Larson, Callie Romero, Aislynn Herrera, Rob, Robert Frank, Anna, Richard Carmen, Kalan Warrick, Shalisa Bynam, and Alex (with apologies for last names we couldn't conjure).

Just as there was the Monte Carlo dynamic trio of friends, James and two other young men formed a similar triad after meeting at LSU. James, Luis, and Doug became a tight unit intertwined with numerous other friends.

Doug reminisces:

There are three memories, particular memories, of James from various times in our friendship. The very first was the day I met James. Luis and I went to Battle of the Bands to watch Adam (Lu's roommate at the time and a high school friend of mine) play. I met James when Luis and I partook in an unprofessional interview of the band. He was visibly annoyed by us and our inane questions on the video, but it was fleeting annoyance as we soon were good friends. I look back at that time now, knowing James, surprised that our interview didn't go over well, but perhaps it had more to do with Got no Tang's placement than the interview itself. There are not many people that I know besides James with whom I could have had such an introduction only to become great friends. The second memory was while I was on a road trip with Kaylee, Rachel, and James. There's only so many times you'll have the chance to pull into an empty desert town, walk into the lobby of the sole motel around, and find the clerk staring into a lone TV on a stand in the middle of the lobby playing nothing but static. If there was ever the feeling you could be taken by hill people in the night, it was then. Nevertheless, the four of us had a grand time that night in White's City. The entire road trip was one great memory, but the uncertainty presented by that town and the humor surrounding the unknown was a highlight. Lastly, I recall giving James a call when I found out Rachel was pregnant and I was going to be a dad. James was the first person I told the news to and he offered to go grab a drink with me at Chimes. We sat at the bar for a long time, me musing about the uncertainty of becoming a dad, he offering words of friendship and encouragement. That drink and his advice were very special gifts I'll never forget.

Luis recalls:

I remember James as being open minded. He was always ready for new adventures, challenges, and experiences. He was very proud. He didn't like being bad at things. He'd insist that no one wait for him when we would go on runs. He didn't much like that he wasn't a good long distance runner. But his pride would motivate him to keep trying. He was kind and understanding. He was an intimate confidant of mine. I could talk to him about my feelings, struggles, and issues. And in that manner, he was compassionate. He was more sensitive than he would have admitted.

From my vantage point as his mother, high school was a segue to ever-evolving college and life opportunities for James. He was eclectic in his learning, which led to a myriad of experiences. These included, but were certainly not limited to: condominium living, a Bachelor of Arts in Anthropology, a minor in Spanish, tutoring, partying, dating, the Brazilian Martial Art of capoeira (a combination of fight and dance), parkour with friends (rapid running, jumping, and climbing through urban areas), disc-jockeying and recording, competing at the International Beard & Moustache Championship complete with Baton Rouge television publicity, bartending at The Chimes near campus, winning the coveted Baton Rouge's Hottest Bartender 2008 award, playing in a band, receiving the Chancellor's Award at LSU, staying on the Dean's List, studying abroad, benefiting from scholarships, and participating in Phi Eta Sigma and the National Society of Collegiate Scholars. James was recognized by family and friends as diverse—striving to be unique and willing to relentlessly work to enhance areas he acknowledged were not his strengths.

He was not innately a particularly versatile athlete. At this juncture, travel back with me to the time of James entering the high school world. Allow me to demonstrate his tenacity in improving himself when he hit what he likely perceived as a wall. In true James style, he pursued athletics in high school. And so followed rugby …

7

Of High School Rugby …
or Of Giving 100%

No eighth-grade student in Brother Martin High School's five-year curriculum ever played rugby for five continuous years … until James attended. As an eighth grader, he asked if he could play. As his parents, we were proud of him for wanting to start early and for subsequently playing five years of rugby through his senior year. I said before that James didn't consider himself an athlete, and yet Coach Gary Giepert encouraged him to the point of excellence in the sport. Gary recalls James wanting to play in eighth grade, a decision he, as coach, had to consider before allowing him to join the high school guys. James didn't know the game, but he was tall and strong.

James played hard and with consistency throughout those years, in spite of shoulder injuries and pain. He won trophies as team captain, best forward, and first player to play (survive!) five years of consistent play. Then, as he began his college career, his injuries required bilateral shoulder surgery. Given the choice, he opted for surgery on both shoulders at the same time. I remember that Bobby Shackleton, our dear friend who is an orthopedist in New Orleans, asked James if he was sure he wanted to do both shoulders at once. And I remember how nervous James was. But James had the gift of healing and recuperating quickly. He was back in no time carrying heavy trays at The Chimes where he worked throughout many of his college semesters.

After James died, Brother Martin High School in New Orleans started a rugby award in his memory to be given each year to students who played five consecutive years of rugby as James was the first to do. For each year ever since, when one or more students have earned this distinction, Jim and I attend the rugby banquet to present this James deBrueys Memorial Award. Some of our children and grandchildren attend and present with us as well.

8

Of Tattoos & Moustaches

James was the king of facial hair. It is a fact that he could grow, cut, regrow, and shape his facial hair and the hair on his head faster than any other person we knew. He could have a beard, moustache, and/or hair on his head one week, not the next, and practically grown again by the next.

Though James participated in international competition for facial hair accolades after he graduated from LSU, he had amused friends and family alike with style for years before that. This entertainment continued until the time he left this earth.

Proud would be one word to describe James' feeling about his Fleur-de-Batman tattoo on his back. When he had this tattoo done, he didn't tell me. He came to our home one day when his siblings were all home. I believe he anticipated that I either wouldn't like his tattoo or at least would be shocked by it. The scene unfolded as James' siblings waited around a corner in the hall and James came into our room to reveal his surprise to me. I responded something like, "Okay! Well, that's nice I guess." I am sorry now to say I think I may have disappointed him. The funny part of this was that there was a collective sigh of disappointment from around the corner in the hall from our other kids. I guess they thought I was that uncool. I'm glad we have a picture of James' Fleur-de-Batman. And I'm honored on behalf of James by all the people, both friends and family, who have special tattoos and facial hair now to remember James.

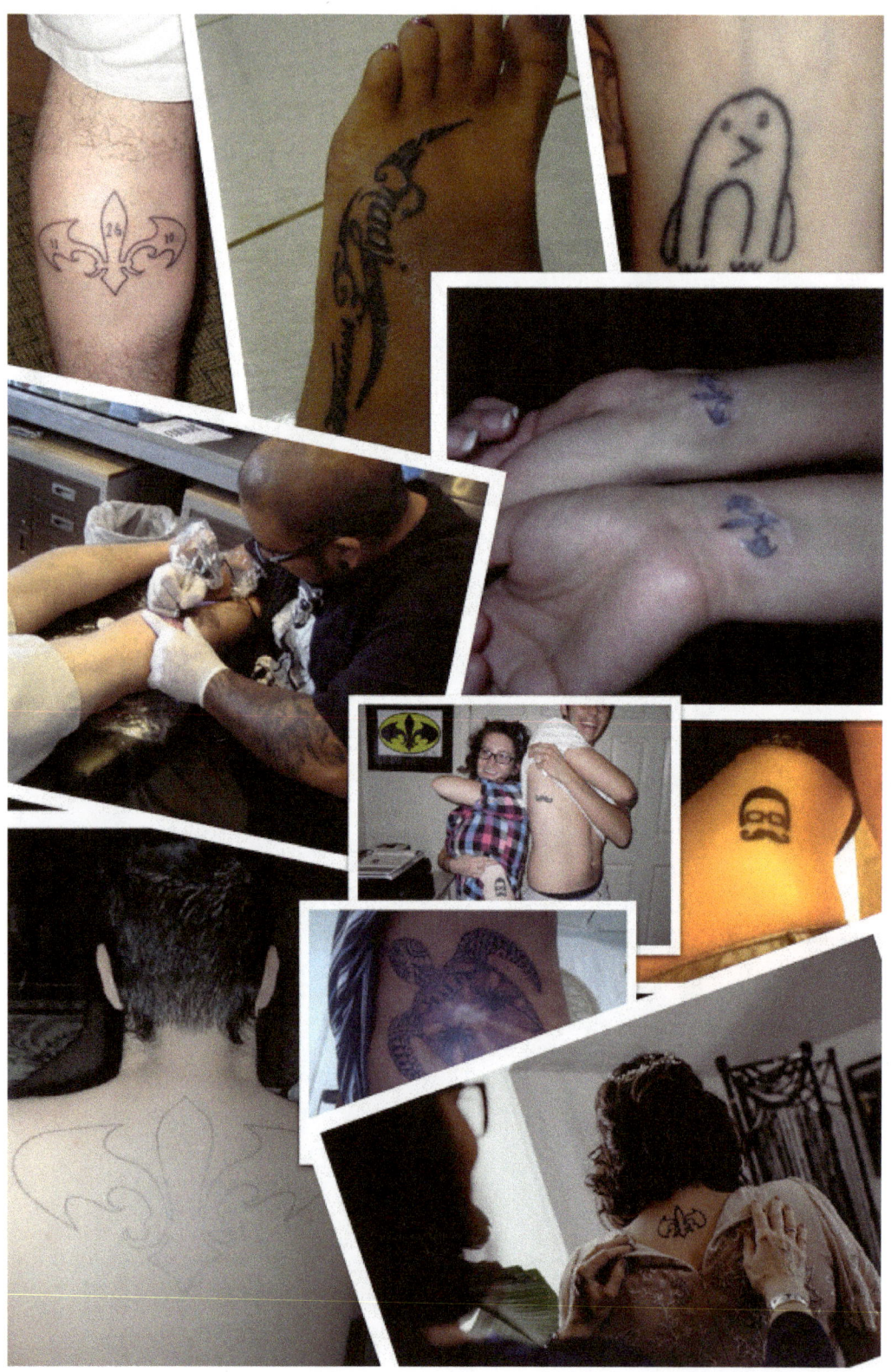

9

Of Dating & Hurricane Katrina

Kaylee and James met in high school, dated, and shared the drama and acclimation process of our post-Katrina world in south Louisiana. Hurricane Katrina hit NOLA in late August 2005. Both James and Kaylee were uprooted from the New Orleans area due to the storm, just as they were embarking on their senior year, James at Brother Martin and Kaylee at Cabrini High School. The first semester of James' senior year was completed through night school on campus at Catholic High School in Baton Rouge, during which time his close friend, Brandon Allen, lived with our family. This allowed them, with Steven, to attend Brother Martin together on the Catholic High campus in the evenings. Since both schools (Catholic High in Baton Rouge and Brother Martin in New Orleans) are run by the Brothers of the Sacred Heart, this arrangement allowed students who stayed in the Baton Rouge area after Hurricane Katrina to be served until the New Orleans campus was sufficiently restored. Kaylee began attending Dutchtown High School in Prairieville, outside Baton Rouge, where she graduated in 2006. James returned to the New Orleans area to live with Brandon for the spring semester in a trailer on the Allens' property while his family continued to recover from tremendous home and property damage. This enabled James and Brandon to return for the final term of their disrupted high school senior year at Brother Martin in order to graduate from their alma mater in 2006. From there, they each attended LSU. Kaylee continued her studies after graduating from her "new" high school. Subsequently, she became a reservist in the Army National Guard and attended LSU. James completed his LSU studies in three and a half years before preparing for his stint with World Teach that would take him to the Marshall Islands.

James and Kaylee were on-again-off-again over this span of years, and they were friends when he left for the RMI. Over the years after James died, Kaylee visited with Jim, me, Simone, and Steven from time to time. We invited her over for sushi or just to visit, and we visited her as she continued to work at The Chimes.

Sadly, we were notified of Kaylee's death in March 2018. In the eulogy at her funeral, we were told that Kaylee had never fully recovered from two significant losses in her young life—those being her struggle to find her place in our post-Katrina world and the death of her "dear friend." Hurricane Katrina was one of the worst storms in history and certainly in our southern part of the United States. The reader can refer to post-Katrina statistics to learn of the years of ongoing recovery that continue even now and to study the statistics related to post-traumatic stress that affected so many. Deaths occurred, property damage was beyond measure, routines were dismantled, and survivors were changed in ways that were not merely physical but deeply emotional. Some seemed to escape unscathed, and others, like our precious Kaylee, were not able to do that. In her young life, the combination of Katrina's aftermath and the death of a significant person proved too much to handle.

JAMBOS WITH JAMES

10

Of Relationships That Endured Time & Even Death

The Monte Carlo neighborhood boys comprised Steven, Chris Vizzini, Brett Bordelon, Caz Hologa, and Jason Papale, who grew up in our pre-Katrina area in Kenner. James' first best friend, Rhett Ward, lived close by and remained a part of the fabric of James' life. A number of other precious friends and their families contributed to the menagerie of those closest to James throughout his younger years and throughout the impressionable years of college life.

Brandon Allen is a dear friend, who shared too many antics to count with James throughout high school and college. Among them was their exploration of Alaska as two competitors in the 2009 International Beard & Moustache Championship. This event occurred the summer after James' December LSU graduation. James shaped his facial hair as a fleur-de-lis, donned a purple suit, and armed himself with Mardi Gras beads to represent Louisiana at the event's parade. Together, he and Brandon (the same young man I mentioned who'd shared families and homes in New Orleans and Baton Rouge post-Katrina) competed, paraded, and partied with hundreds if not thousands of others from around the world who prided themselves on their beards and moustaches. They mingled there with representatives of many countries, proudly answering questions about Louisiana culture, New Orleans, and particularly Mardi Gras for all who were interested. James won fourth place in his division of competition—as did every other competitor who didn't place first, second, or third.

Emily Eck and James dated on and off over a couple years. They shared a love of music, their time on staff at KLSU, and a core group of close friends. Emily later wrote to me, "I think James and I may've first met when a group of us gathered on campus to play capture the flag in the woods."

During his last semester at LSU, James continued to bartend at the well-known, vintage establishment known as The Chimes. Here, professors, students, and families alike mingled at the bar and had meals on the edge of campus. Some of James' friends nominated him for a city-wide contest to identify the Hottest

Bartender of Baton Rouge. He had so many friends line up to vote for him that he won. What a distinction! As parents we thought it was fun and even funny, though we didn't attend the award ceremony … which as I recall was fine with James. Simone and Steven were there to see him receive his accolade.

JAMBOS WITH JAMES

11

Of James' Resolve to Stay In Touch & Tie Up Loose Ends

James was excited to go to the Republic of Marshall Islands. He taught English at Notre Dame Seminary in New Orleans in preparation for working with non-English speakers. He filled out papers, completed the application process, and got the very few things he needed. His priority seemed to be to solidify his plans to maintain contacts with friends and family while he was on the other side of the planet.

As he got closer to the day of departure, he demonstrated this deep-seated priority to maintain contact with friends through a number of initiatives. He gathered addresses and contact information, recognizing that contact would be a challenge living in such a remote, unplugged part of the world. A key avenue might be social media if he was afforded the opportunity, such as demonstrated by this post on Facebook. At that point in time, Facebook was a fairly new communication tool that was to be such a gift to him and us over the little bit of time we unknowingly had left to share with him.

Here is James' entry shortly before he left for the RMI.

James de Brueys
2 July 2010 ·
Contact in the Marshall Islands
First off, this is the address:

James de Brueys
c/o WorldTeach
P.O. Box 627
Majuro, MH 96960
Republic of the Marshall Islands

*International rates for postage may apply ... I'm not sure. Priority mail moves the quickest, and if you send a package send flat rate boxes. If you send a package, it **MUST** have a customs form (ask for one at the post office).*

I will be closer to the main island than not, so the mail may not take that long to reach me. However, I have heard different stories about mail taking anywhere from one week to 6 months ... so don't send me any guava or ferrets. They'll probably both die.

When I am on the main island, I may have some access to the netternet. I have a Skype account: the name is JamesCdeB. I will try to write letters out, but it may be a little more difficult, as I would have to just hand it to someone on a boat most of the time and ask them to deliver it to the post office on the main island (Majuro).

Just an FYI, I will be teaching at the Bikarej/Bikarej Elementary School on the Arno Atoll. Arno is the closest atoll to Majuro. Look at it on Google Maps. You can zoom out to get a better perspective of where I will be.

And that is about it right there! I hope to hear from all of yall in the future! I leave N.O. July 21st, and will be in the Marshalls on the 23rd of July.

<3

James

P.S. I guess don't send me coconuts either... I think I may have one or two while I am there.

***

James had decided before he left to ask friends to take care of certain of his personal belongings for safe-keeping. He owned very little, a life-style that seemed to suit him. He had friends keep such possessions as his flute, a sword, and his bike. We later decided that the items left for safe-keeping with his friends would remain with those friends to whom they were entrusted and who treasured them.

James left certain keepsakes with us and departed for the Marshall Islands with few clothes, a camera, footwear for walking on coral, a solar charger, an old phone he knew he would rarely have a chance to use, writing supplies, and a Kindle loaded with as many books as he could collect and fit. (When James had been in the Marshalls only a short time, his solar charger landed in the ocean and was ruined, rendering many of his possessions useless.)

Yokwe James,

Congratulations! You have been selected to teach at Bikarej Elementary School on Arno Atoll. You are going to be the first WorldTeach teacher to be placed there, so I do not have a handover letter for you. However, when I was a volunteer (back in 2006 on Ulien, Arno), Bikarej was one of the close by islands. I can tell you that you have about 60 students in the school, and while you are in the atoll closest to Majuro, Bikarej is located in the far north, actually in between the big, main Arno lagoon, and a little tiny one- which is very unique and cool! Also, Bikarej generally has their own boat, which either goes back and forth to Majuro, or picks people and food up in Arno, Arno, which is where teh main taxi boat goes. Attached is the handing over letter from our most recent volunteer on Ulien, Arno. While it isn't the same site, it will still provide you with some general information about the atoll. Please feel free to contact me with any further questions and once again, congratulations!

We look forward to meeting you soon!

Annie and Angela

--

Annie Himmelsteib
Field Director, Marshall Islands

WorldTeach
PO Box 627
Majuro, MH 96960
MARSHALL ISLANDS

12

Of World Teach & Marshall Islands Life

James left for the Marshalls on July 21, 2010. From New Orleans, he traveled to meet a number of volunteers in California, who then together traveled to the RMI. Jim and I took him to the NOLA airport and snapped the one last photo shown in the collage within this section. We had just hugged each other and cried. Looking at this picture, I have always felt that James looked sad. For all of his adventurous spirit, I imagine he had some misgivings about leaving home and everyone in Louisiana for such a remote location.

Also included in the collage is a remarkable photo of James' World Teach group. They all were assigned to the Republic of the Marshall Islands and many of them had just met that day in California to make the trans-Atlantic trip to their newly assigned home for the coming year (note: letter dated 08/06/10). I am unutterably grateful for these last friends with whom James had contact before his death.

***

On September 10, 2010, approximately two months into James' time in the RMI, I posted for family and friends on Facebook about my particular joy over having communicated with each of our five children on that specific day—though they were worlds apart geographically. This was written shortly after James left Louisiana and two months and fifteen days before James and the others with him were lost at sea. My post read:

> "Talked today with André and John who are in Spain, 'chatted' with James from the main island (Marshall Islands)—an unexpected treat since he is there on an unannounced (at least to us!) long weekend, talked with Steven in Atlanta and picking him up tomorrow from the airport for a 3-day weekend with us, and blessed with Simone and "Michelle et al." every day in LA with us! Life is good!"

I also found this post from James' sister, André, two days before he died. Being so unplugged there, I wonder if he ever saw it …

Her post included a photo and said:

"Got your letter today at work, and it made my day. It sounds like you are keeping the Marshallese folks on their toes. After reading your letter, I was looking at some pictures and I found this one, and it made me miss you ☹. Are you hairy again? Can't wait til you are in some sort of 21st century civilization so we can talk to you. We will be sending some mail your way very soon. Love you!!!!! Dré

***

Of all the chapters in *Jambos*, I believe writing this section of the book is one of my greatest challenges. Photos may be the second purest means of understanding James' job as a volunteer and the joy of the developing relationships James experienced. His letters are our sharp instrument of insight. His World Teach and RMI contacts became family to James; they were the last people he saw and with whom he communicated in his young, soon-to-be-ended life. I value them more than I can well communicate.

In the large photo that follows this section, you see James and his volunteer group in Majuro. Majuro is the capital of the RMI. You can appreciate the awesome Pacific Ocean in the background. From top row left to right to the bottom row left to right are: Jordan Rodgers, Angelina Jimenez, Todd Mulroy, Lara Farina, Kempton Baldridge, Ken Hagberg, our James, Brooke Payne, Stephen Leard, Hanneke Van Dyke, Max Hoegh, Laura Sunblad, Annie Himmelsteib, Angela Saunders, Miranda Blount, Kristina Bramwell, Katie Finberg, Mary Peters, Sarah Leard, Belinda Fenner, Joann Dai, Mandy Doyle, Lindsey Ryan, Clare Jones, Valerie Lazickas, Hannah Thornton, Stephanie McCutcheon, Noa Silva, Amanda Colon, Karen Nugent, Aryn Kamerer, Christina Marie Stenhouse, Monica Montgomery, and Erica Moore.

Our family met most of these amazing individuals in the Marshall Islands when we traveled there very quickly after James' boat went missing. We attended memorials on the main island in Majuro and on Arno Atoll in Bikarej where James lived. These were celebrations and opportunities to visit the individuals James had been getting to know. Some of the World Teach team members were not able to join us for the services because of insufficient transportation to bring them in to the main island from their remote locations.

We have been told that it is not unusual for a number of volunteers to leave their assignments each year in such positions as James and his fellow volunteers had. The work and conditions are hard—challenging due to culture, language, and absence of daily amenities. Only two left their posts later in the year that James was assigned to the RMI. This group was unique and compassionate.

Some RMI atolls had running water and electricity. In Bikarej, James had no electricity, no running water, no roads, and no cars. His village had a generator that had been broken since long before James arrived. He had a one-room school house with little to no paper, books, or any materials. There was no plumbing. In fact the school actually had not graduated any student for years. James was the first "*ribelle*" (white man) to ever volunteer in that location. The reader will see that James refers to some of these distinctions throughout his writing.

Of these young volunteers who knew James, roughly two-thirds have maintained contact with us over these past ten years. Half of those continue to share their lives, growing families, and jobs with us personally and regularly. A dozen have made trips to Baton Rouge and New Orleans to meet us again or for the first time. We believe James found an incredibly compatible group of friends in his brief stint halfway around the world. As a group, bonded together by their work, and bonded then perhaps more tightly still by the loss of one of their own, their gentle spirits, love of writing letters, and service-based natures like James' were nurtured. James was blessed ... and so have we been blessed ... by these young adults.

James' letters tell his story of his students, his teaching, living on an island/atoll with less than 200 people in all, the culture (letter of 08/27/10), new friends (letter of 08/06/10), patience living on Marshallese time (letter of 10/16/10), the Baha'i faith (letter of 08/29/10), TEFL certification (letter of 11/04/10), copra (letter of 08/28/10), sharks (letter of 08/27/10), little to no privacy (letter of 11/04/10), and the two families—his island families—who adopted him (letter of 08/20/10).

The parents in a host home are called that volunteer's *Mama* and *Baba*. Thus volunteers assigned to relatives' homes were considered James' cousins. Photos, sent to us in great part by these families and new friends in the RMI, serve as a complement to James' letters. They highlight their comradery and relationships as new friends, who shared the challenges and uniqueness of their remote and unique island world.

Among the greatest gifts we have from James are his writing and our photos that authenticate the story of his last few months of life unfolding over 6,000 miles away from us.

***

Helen Claire Sievers was the World Teach Director when James was assigned to the Marshall Islands. My family and I are grateful for her sympathy and support throughout our time of learning of James' boat having gone missing. She warmly helped us work through the details of travel so that our family could quickly and unexpectedly make plans to journey to the RMI. Helen Claire stayed in close touch with us throughout this process. I remember being grateful when she mailed us some of James' handwritten lesson plans that she had in her office (note: letter of 08/20/10 and letter of 08/29/10).

Annie Himmelsteib and Angela Saunders were co-leaders of the World Teach group of which James was a part. Annie and Angela had served World Teach prior, and they had moved into positions as trainers and directors for James' group.

Angela stayed with World Teach until mid-2011, following which she became a graduate student. From 2013 until the present, she has served with the International Organization for Migration. She has had several titles and at the time of this writing serves as the IOM Marshall Islands Head of Sub Office and is the UN Joint Presence Office Coordinator. Angela wrote such compassionate words to us after James and the others on the boat with him died: *"Getting to know James during orientation and throughout the year during radio check-in, in letters, or when he would come to the radio to ask a question or relay a funny story or message was truly a privilege. It was clear from the beginning that James acted as a type of glue that helped to hold our World Teach group together and make it whole."*

When the search ended, Annie wrote, "He IS personality—everyone was drawn to him—the other volunteers during orientation, his students, and family on Bikarej. I am so glad to have known him and seen him thrive here in the Marshalls. I will always think of him as the two-syllable Jam-es that he is known by throughout Arno—always said with love and admiration."

The volunteers compiled a book of photos and notes to James to present to us when we arrived in Majuro. Later, they sent a package with copies and even some originals of many letters collected from the volunteers that James had written to them. Though these are not referenced or copied here, they are another chapter of James' story yet to be preserved. Volunteer-to-volunteer letters have even still a different flavor and insight than all I am sharing here. Perhaps a sequel …

Erica in the RMI wrote at James' memorial, "*My fellow Arno-ite. I remember meeting a group of Bikarej women in October and the huge smiles that were on all of our faces as we talked about you. You were so happy and in love with life. The last time I saw you I promised I would bring your positive energy and attitude back with me to Arno.*"

Brooke at that same time wrote, "*James helped me form a new attitude about my time here—something I'll be forever grateful for. Kommol Tata Jam-es, Brooke.*"

Mary added, "*You have made an incredible impact on my life. During orientation we bonded over the great, classic movie The Newsies as well as how great our families are. You had the rare gift of making everyone feel included and loved. I will always think about your impressive coconut husking skills during orientation. Thinking of you, Mary. 'Open the gates and seize the day …'*"

Christine wrote, "*Thank you for sharing your life with me. I will always admire your selfless service and good works. Your love for God was evident. Your light shines on in the Marshall Islands.*"

When James won the husk-off during the July 2010 training for cracking into a coconut the fastest, (maybe because he used to shuck oysters where he worked back in Louisiana?), we are told he remarked, "I'll tell you who the winner of this was: everyone. Because everyone wins in a husk-off!" I can hear him proclaiming that in true James form.

Clare Jones wrote kindly to us then, and she continues more remarkably to write to us through lovely, hand-written, actual posted mail ever since. She sends art and ballet clippings to me; keeps us posted on her radio shows; makes sure we are current on events within her family, since she was the one other Louisiana World Teach volunteer sent to the RMI; and often attends our Open Homes each year that we started celebrating for James' friends and family at the end of November. (These Open Homes have now expanded to a gathering of all family and friends who care to join us.) In December 2018, in communicating with me about the potential publishing of James' letters, Clare wrote, "*I was so pleased you were able to come by on Saturday* [to her birthday party in NOLA]. *I would love to be involved (in whatever capacity is most helpful) in your epistolary project in James' memory—letters are a lovely legacy that writers, artists, and just wonderful humans leave behind. They tell stories all their own, in their own words.*" I am grateful to Clare and her parents, Serena and Kirk, for maintained check-ins as well as moral support for *Jambos With James*.

Annie and Angela had contact with James before he went to the Marshalls. They guided him in his work, encouraged him tremendously through his reviews of his work there, and were considered by James to be friends. Sadly, as co-leaders for World Teach, they had the daunting task of staying in touch with us during the Coast Guard search. I remember thinking, even while it was going on and certainly after the search ended, how exceedingly sad and difficult this unwanted and unpredicted responsibility within their job descriptions must have been for them. They then had to be superglue for their group who had lost a fellow volunteer. They had to continue their work and service under the dark cloud of the deaths of five people on that boat. We are grateful that Angela and Annie were there for our family during our visit to the Marshalls in December 2010, and that they have stayed in touch with us over the past ten years.

Our rock and go-to person for RMI contact, from that first year through the present, has been Todd Mulroy. He was an outstanding member of James' volunteer group, who remained in service to World Teach through July 2015 after his initial volunteer year. Two other jobs since that time continue to keep Todd tied to the Marshall Islands. Todd befriended our entire family from the beginning. Though we did not meet him at the memorials, since he was one of the volunteers who could not get transportation to make it to the main island from his assigned atoll, it always felt like we knew Todd personally. His willingness to share information and to help us maintain contacts in the RMI has been above and beyond the call of duty. We have never yet been able to get him to Louisiana … but not for lack of trying that continues.

Since we have corresponded so often with Todd, imagine my surprise in December 2016, when texting with him that I first consciously realized that Jim and I had never actually met Todd in person. This realization stopped me in my tracks. I couldn't believe it was true—but in fact it was. We have been in such close contact with each other throughout this journey, but we have never met. We all agree that a meeting in person will still happen.

Here are some samples of the texting back and forth maintained with Todd that kept us connected.

This first entry is one back-and-forth with Todd in early 2016:

M. Awesome, Todd! I just came back to my computer, and I'm so glad I did! Are the first people I see James' island mom and his brother? I am so happy to see "Baby" James and his dad … how heartwarming. Todd, you are the best!

T. Yes, Emita is James host mama. Baku is his host brother. He is in the picture before with Salem. His real name is David, but they call him baku.

M. And her husband was the captain on James' and the others' boat that was lost, correct? Little James is adorable, by the way!

T. Correct. He's so cute. He ran up to me and gave me a big hug as I walked in the door. I had no idea he remembered me.

M. What is the other name on the box he's holding. Looks like Bamle maybe?

T. I wrote on the boxes who they're to and from—Mary T deBrueys and Baamle." Baamle means family.

M. Do you think little James sort of gets who you and we are to him? I guess his dad knows, so hopefully he will explain more as James is older. I hope we can stay in touch with him somehow over time!

T. I think he gets it. I am sure it's spoken about often. They know when they see me it's somehow related to James.

M. That's wonderful!

T. Some older boys on Bikarej call me 'James' friend'. They don't know my real name.

M. Aww! Todd, what is the RMI going to ever do without you whenever you do come back to the States? (… assuming you do come back eventually!)

T. Haha. Yeah. I don't know. It's hard for me to even think about. Part of me never wants to leave. But in time I'll have a hard decision to make.

M. I can only imagine how hard it will be. One thing we still hope for, though, is a chance to visit with you here eventually! (I didn't fully realize we had never met Todd in person at this point.)

T. I can promise you we will eventually meet. Seeing Steven was hard enough. I can't wait to meet the rest of you.

M. We will welcome you with open arms, feed you, and take you wherever you want to go when you visit! Looking so forward to it! Good night! Have a good weekend!

T. That's super sweet. I can't wait. Good night.

And so followed another conversation online in December 2016:

M. Todd, Jim and I have a quick question for you! Have we ever met you in person? Jim and I feel like we have, but you said something in a recent message about not meeting us that made us wonder! It certainly seems like we know you from meeting you, but I guess it's possible that our contact has always been only on facebook, by email, and in the mail! If that's the case, wow!

T. We never met in person! I have only met Steven in person.

M. I guess it seems like we would have met you when we all traveled to the Marshalls. Since we have had so much contact with you, in my mind it feels like we've always known you. I'll let Jim know that you responded to my message. And, at any rate, we can't wait to eventually meet you whenever that happens!

T. When you were in the Marshalls I had no transport to Majuro so I missed the gathering. I know for sure we will eventually meet up.

M. Todd, where were you originally stationed in the RMI?

T. In Aur, Aur.

M. Got it. Thanks!

T. Laura Sundblad was on Tobal, Aur and she had the airport. We had no fuel on my island for me to get to Tobal to catch the plane.

M. I recall that some people couldn't get in to Majuro. I think because of the photos, like the one of the whole group where I can clearly see you, I just feel like you were there. I guess we felt your spirit even before we knew about you!

T. I was devastated. I wanted to be there so bad. Ever since they told me that rescue and search boats were coming my way, I walked my whole island circumference every day for 3 weeks waiting and looking. I felt obligated to tell you that in Majuro but I never got in. I wrote you a letter but was never sure if you ever got it as other pieces of mail I had with it got lost in delivery. But my heart tells me we will meet when I leave this place.

Thanks so much for the kind and sort of happy/sad note, Todd. I just showed it to Mr. Jim, too. I don't think we ever got that letter you mentioned from 6 years ago. It seems like one I would remember, though I must say I am pretty sure my memory may not have been 100% during that very difficult time. I do still *have everything everyone sent back then, though. In any case, I am actually glad to hear this from you now, and the chance is much better that it will stay impressed in my mind. We look so forward to having you here for a visit whenever that is meant to happen!*

In July 2019, Todd wrote, *"I was in Arno, Arno for a few days and met up with Amitha, Salem and Ducky. Ducky was 4 when Steven came for the dedication, and he is now about to turn 12. I asked about baby James and she (Amitha) said he is in Bikarej. She will give a message that we are looking for him.*

***

All of the kindnesses, notes, and photos of James that his peers sent to us from his time in the RMI mean the world to us. As parents, we are thankful for all James experienced in his precious four months there. He expressed in his letters and in the few actual conversations we were able to have with him when he was on the main island that he was happy, peaceful, and had wonderful time for reflection.

We are grateful.

JAMBOS WITH JAMES

13

Of First Responders in the RMI—The Coast Guard & the Pacific Effort

Reflecting on James and the past ten years, Jim (James' dad) wrote:

Driving to Arkansas to watch the LSU Tigers play the Razorbacks. A trip planned by Mary T and me over the course of several weeks for a fun weekend get-away at a B&B that was going to be a respite from a busy semester. We were looking forward to a relaxing weekend with good food, football, and each other. Getting off the highway to get to our Little Rock B&B, we were interrupted by a call on my cell. It was Thanksgiving week, so I thought it was probably just one of our kids checking in with us.

Voice on the phone was one I didn't recognize. "Mr. deBrueys? This is (name), Charge' d'Affaires at the US Embassy in the Marshall Islands. We're calling to let you know that a boat transporting your son, James, and others has gone missing. There is a search in progress and the US Coast Guard has been notified."

This was our first notification that James was lost at sea. Ten years ago and it seems like yesterday.

***

Our world stopped.

We couldn't think clearly. Should we turn around and head back to Baton Rouge? Should we sleep first since we had just driven 350 miles? Could we sleep at all? What could we do from over 6,000 miles away? Was James dead? Was there a chance he was alive? How long can one survive in the ocean? What if … Our imaginations played out possible scenarios rapidly and yet in sort of slow motion.

We decided to check in at the B&B where we had a room, to try to sleep before turning around to go back to Louisiana. I don't remember what, where, or if we ate. I remember that Jim lay down on the bed and dozed off and on and that I couldn't. I prayed that the Blessed Mother would take care of my baby boy as a mother would - as I wished I could. I felt like I knew then that James had died. I became aware that I wasn't praying that he was alive … just that she would take care of him. Like maybe he was already in Heaven. I prayed he was in Heaven. I walked out on the balcony of our room (second floor, I think) and called my friend, Carolyn. We talked. Did I pray that his body would be found? I don't remember. I wanted that, but I don't recall if it was a prayer.

Before the sun came up and before breakfast preparation began downstairs, we left the B&B to head back to Baton Rouge. We penned the B&B folks a note because no one was available that early. I remember thinking we should have left the LSU tickets for someone to use. I don't remember noticing over the next day or so if LSU won or not.

As an aside, later that week we learned that the B&B charged us for a second night because we didn't give 24-hour notice, though we left them a note that we had just received word that our son was lost at sea in the Pacific and so we had to leave immediately. I called them later, too. I felt angry at them for their ruling. But I also didn't care enough to pursue it. They later let us know we could come back for a two-night stay another time and would only have to pay for one night. I remember thinking, *No, thank you.*

Meanwhile there were constant calls (sometimes not seeming like they were frequent enough; sometimes feeling like they came without time for us to catch our collective breath between them) from the RMI and the Coast Guard and the US Embassy … on and on. We were notified of the efforts—no sightings yet—and the boat not being reported missing for a whole day—and that the boat had left on Thanksgiving but that the Marshallese are often so off-schedule that no one reports much until some time has passed—and that there was maybe a sighting—and that that was a false reporting—and that some but not all were found—and that the boat was found bobbing in the ocean but with no passengers—and that the boat was small for so many people—and that an ice chest of fish was found intact—etc., etc., etc. We began to refer to time in the RMI as "Marshallese time," as did James in his letters. Things just happen when they happen; schedules change on a dime (note: letter dated 09/11/10). We arrived back in Louisiana from Arkansas and first stopped in the Northshore area of New Orleans to tell Jim's dad what had happened and what was going on from what we knew. It all didn't seem real. Jim's dad went into another room and cried. Then he came back and said he knew God was with us and felt assured James was okay (Here on earth? In heaven? I wondered). I believed him; I wanted to believe him.

More calls from the Pacific. Still looking. The Australian Coast Guard was helping, too. Everyone on the islands with boats was out looking locally while both Coast Guards were surveying the vast expanse of ocean around the Marshalls.

My mom had been living with us shortly before all of this happened. We were remodeling a part of our home to include a roll-in shower and easier access for her wheelchair. During that time, she had gone to stay with my sister, Paula, and her family. So our next stop, coming back from Arkansas, was at Paula's home in Baton Rouge to tell my mom and Paula what was going on. I can still see myself walking down the couple of steps into the sunken living room to talk with them. I remember it again felt surreal—standing in a living room talking to my mom as we had with Jim's dad. Mom looked like she was trying to understand, to take it all in.

Then we went home. To our home. Our surreally quiet home. We waited for the phone to ring over and over for those next few days. Most calls came at hours opposite our waking hours since the Marshalls are at the other side of the world.

We contacted each of our children. Steven was doing an internship in Georgia. We didn't want him to drive home alone. Jim flew up and they drove home together. André and her husband, John, were in Spain teaching English. They flew home. This was their second trip home for a death. During a previous stint of theirs in England, John's sister, his only sibling, had died of cancer. This was just shortly before James died. Could this all be really happening? Michelle, her husband, Pat, and their two oldest who were born by that time, Madison and Blake, were in Kenner at their home. Simone was still with us in Baton Rouge. We would quickly have all our children physically home … all but one of them. Word spread and help came from so many of our family and friends. Prayers, enrollments, cards, calls, visits, food—it all came.

Around this time, we were introduced to names of various responders, including Mark Morin from the US Coast Guard. We became acquainted with words and terms like: the US Coast Guard 14th District, the Coast Guard Air Station of Barber's Point, the Department of the Navy, and the Marshall Islands Marine Resource Authority. These parties were involved in the search; they were using a USCG HC-130 Hercules in the search, from an air station in Barber's Point, Hawaii; there was a US Navy C-130 from Philadelphia joining the effort; and other patrol vessels, fisheries boats, and buoys were at play. We learned that the distance from Majuro in the Marshalls to Oahu, Hawaii was 1,975 nautical miles. A number of trips searching the wide expanse of ocean in the area were made. The report from the Coast Guard indicated that visual departure of James' boat had occurred at 5:00 in the evening on Thursday, November 25, 2010 HST. The Coast Guard was notified by the RMI at 4:20 in the evening on Friday (24 hours later) about what was referred to as the "non-arrival of the boat from the Marshalls". The USCG launched their investigation officially on Saturday, November 27th. We learned of Alpha Searches on November 27th, Bravo Searches on November 28th, Charlie Searches on November 29th, Delta Searches on November 30th, and Echo Searches on December 1st. We also learned, finally, of a reference to a Foxtrot Search on December 2nd. Our recollection is that this last one was due to our request/begging for more search time, reinforced by requests from some supportive State Representatives from Louisiana. The total area searched was 12,512 square nautical miles. The total time invested in the search was 128 hours. The total combined count of sorties was nineteen. (A sortie is defined as a rapid movement of troops or a sudden issue of troops.) On December 1st we were told preliminarily that the search was to be called off; on December 2nd after the granted extension, the end of the search was finally confirmed. We were also later told by the Coast Guard when we met with them in Hawaii en route to the Marshalls that it is not possible for a human being to live in ocean water for more than a couple days. References told us that, for reasons not completely understood, human skin begins to break down in water after a few days. These are visuals and words that no parent—no person—wants ever to hear.

Hearing the words "THE SEARCH IS CALLED OFF" was probably the saddest and most sobering utterance we would ever hear as a mom and dad. We had been granted so much search time, likely quite a bit more than we reasonably should have had, but those words still seemed so harsh and premature … and so unfathomably final.

14

Of First Responders in Louisiana

So much was happening so fast. For a few days, our time was consumed with getting all of our other children home and sitting by the phone. Calls from the Embassy, the Coast Guard, and World Teach were intermittent and at any time of day or night, mostly during the evening and night hours. This made our days seem unbearably long.

A priority of preparing to travel reared its head suddenly. We needed airline tickets and we needed passports updated or, for most of us, for the first time. Passports take time to process. We had to hurry to get there for memorials and to visit where James had been living to procure his few but now so precious belongings. My mother had always been stellar in her willingness to jump in when times were tough. She was immediately available and stayed present for us in the same manner she had for our children and me when Jim had suffered an aneurysm ten years earlier. Now, with my mom's dedication and help and that of my sister Paula, passports were rushed through the proper office in New Orleans. We went there on short notice, and my mom paid for all rush fees to get this done. Paula executed the details. I didn't worry about any of it or organize anything. We just did what we were told. Before we knew it, the search for James and the others on their boat in the Pacific had ended, and Jim and I with James' four adult siblings were en route to the Marshall Islands - a place that we had only ever heard of and had come to learn a bit about within the last year. We were so grateful at the time that our two sons-in-law so willingly stayed behind to take care of all at home.

Several remarkable prayer warriors were on our team. I knew always that Carolyn, Sally, Marilyn, Debbie and her sister (Mary Ann, Adam's aunt), Mary Tebo, Jeanne, Susan, and others were on their knees for James and all of us. I also had a special team of family and friends who came to our home from time to time and just sat with me. They helped when we needed help and were otherwise quiet unless we wanted to talk. They were among my greatest blessings in this new world of utter grief and shock. Thanks are due to Paula; Beth (my friend since second grade); Mary Alice (my friend since first grade); Deb (my friend since kindergarten); Carolyn (my co-worker for twenty-five years and friend for longer); Barbara (my friend since second grade); and Kim (my friend since our freshman year of high school). I called them my assembly of angels at that time. They were with us, just with us.

Another of our first responders in Baton Rouge was Cary Messina. A friend of mine from elementary school, Jolie Ranzino, is married to Cary. Together they were so generous to our family, with Cary jumping in to work through the succession process for James' meager "estate" at no cost to us. $1,969.06—this was the total estate value of James' possessions for his small succession.

Also among our amazing first responders in Louisiana was a now-friend we met because of James' disappearance. Miranda Levi-Strauss Powell, a young Marshallese lady living in New Orleans, contacted us to let us know that she was available to help in any way she could. She is married to Ryan Powell, a scout for the New Orleans Saints football team. Born and raised in her earliest years in the RMI, Miranda attended school and university in the United States, during which time she met Ryan. Together they have three beautiful children, a daughter and two sons. We have been gifted with a friendship with this family as a result of our loss. Miranda was used to traveling to and from the RMI, and she offered to facilitate our trip to the Marshalls and to attend the memorial on the main island. She also connected us with her mom and dad, Mona Levy-Strauss and David Strauss, in the Marshall Islands. They met us there upon arrival in Majuro, as did many of the World Teach volunteers. I remember the Strausses gifting us with beautiful and unique *amimono* (crafts made in the Marshalls from coconut and pandanus plants, often with incredible, local seashells). I remember them helping us navigate the atoll. I remember them gathering newspapers for us to have records from the Marshallese perspective of our loss. They fed us and were there for us. Since that time, we have seen the Strausses in Louisiana a number of times for meals and warm, entertaining visits, often including the Powells as well. How incredible!

One of James' friends who immediately became a right hand for us was Jess Smith. Jess was a student in law school at the time, and she and James were tight friends. Jess offered to help with the memorial and anything else we needed. I remember that we grabbed her in the parking lot at St. Aloysius before our memorial Mass on December 9, 2010 to ask her to field a couple of reporters who were waiting for us in the parking lot when we arrived. She took that task with such composure and grace, and I remember being so grateful for her at that moment.

Several of James' closest friends orchestrated a gathering of friends and family on December 10, 2010 at The Chimes. Located at the edge of the LSU campus, this old establishment has become a favorite of Jim's and mine again so many years after our own college days at LSU. Allyce Baudier Lemon and Brady Lemon had a huge banner made so that James' friends and family could sign it. On the banner was our "Bring James Home" logo designed by Ali Becnel Solino, another of James' friends, who is married to Adam Solino. Nikki Fuchs Vidrine made and framed an incredibly on-target collage for us, capturing many facets of James, and J.J. Alcantera penned and framed the Fleur-de-Batman that James had created and for which he had become known. One of Michelle's closest lifelong friends, Caroline Vanek, gave us many photos from the day at The Chimes, and André helped us integrate them into a photo book. A friend who taught with André at Our Lady of Mercy in Baton Rouge, Rachelle Berkley, and her husband, Josh, mass-produced t-shirts for family and friends to be used to raise funds for the basketball court on Arno Atoll. Revisiting all of these kindnesses now, I realize how many of our other children's friends were there, too, to help support them, in turn supporting our entire family.

Prior to the Chimes event, some of James' dear friends got together to establish an effort to fund any attempt possible to "bring James home," including their desire to help our family go to the Marshalls in the aftermath. They raised money for these priorities through events such as the one at The Chimes and through concomitant t-shirt sales. After some dust settled, it was decided to use the funds to build the memorial basketball court in Bikarej (note: letter dated 09/25/10).

15

Of Learning About Things We Never Wanted to Know

There are the things that are hard to see and impossible to un-see.

Pictured within these pages is the small boat that carried four adults, one being the woman with her unborn child, on the day our precious loved ones were lost at sea. It was never fully determined if the boat had mechanical trouble or if the accident was caused by a storm. A combination of those may have been at play. For a long while, I couldn't look at the photo of the boat, which was eventually recovered and brought to the dock in Majuro. An intact ice chest filled with fish was also recovered from the ocean, as was the body of one other young man from the boat. The young lady who was pregnant had been sighted, but when recovery was attempted her body could no longer be found. James' body and that of his host *Baba*, the captain of the boat, were never located.

Because James was an American citizen in international waters, the death certificate process was tough to complete. It took more time than a death certificate otherwise would have, which in turn caused a challenge with getting an obituary in the newspaper in New Orleans. We were told by the newspaper that because historically some people have played jokes on others by ordering false obituaries, one can no longer be obtained without a valid death certificate. So, in New Orleans, we actually had to do the equivalent of taking out a newspaper ad. We then requested it be put on a page next to the obituaries in order to get information about James into the paper while we waited for the CRODA. Gratefully, we posted with more ease in the Baton Rouge paper.

And what is a CRODA? It is a Consular Report of Death of a U.S. Citizen Abroad, issued by the U.S. Embassy or Consulate upon finding of death by a competent local authority. This administrative document provides essential facts about the death and disposition of remains. Our disposition-of-remains line says, "lost at sea." Our cause of death for James reads, "lost at sea; presumed to have drowned." The date of notification was 11/27/2010; the date of the copy of the report sent was 01/04/2011. Interestingly, the notification to the Social Security Administration had been checked by the time of this report, but the SSA told me in the spring that their records showed James still alive. It's funny what you remember from such tragic situations. The death certificate issued by the Marshall Islands says Arno 2010-8. I wonder if that means James was only the eighth person to die in the RMI in 2010. I never asked.

We learned there had been no life vests on the boat. Life vests were not required for travel in the Marshall Islands and were not required of the volunteers by World Teach. I remember thinking of the analogy of: car/seatbelts; boat/life vests. Right?

16

Of Memorials

Before our Baton Rouge Memorial Mass was scheduled, we were unsure about the status of the search for what seemed like such a long time. During that period of grueling uncertainty, two dear friends asked if we would like for them to plan a prayer vigil at our parish church. Beth Adler Nesom and Cathy Selleck Carmouche began preparation for this prayer offering with the help of one of our deacons. Sadly, while the vigil was being planned, the called-off search was announced and plans for our vigil turned into plans for James' memorial.

In Baton Rouge, our December 9th memorial was so personal, special, and attended by many friends and family. I remember my mom and Jim's dad, James' grandparents, sitting at the front of the church and looking so frail. Someone brought me a beautiful, red-beaded rosary that I clung to, and to this day I can't recall who gave it to me. Many flowers, much food, lots of cards, and prayer memorials were offered, and so many kind faces waited patiently in line to offer their condolences.

In a processional to begin the service, James' friends carried photos of James, one from each year of his life. Eric McCrary, Michelle's brother-in-law, served as cantor; Ken Thevenet was our organist. Fr. Gerald Burns and Fr. Donald Blanchard celebrated the service so awesomely. They were joined by Fr. Howard Hall, a friend of our family and in particular of my brother, Kevin, who had died in 1985. Our deacon celebrating with us was Dan Cordes; his wife, Nat, also joined us. Dan had been James' Confirmation sponsor when he was younger. Concelebrants who honored us with their presence were Fr. Jose' Lavastida, a friend from New Orleans, and Brother Ray Hebert from Catholic High in Baton Rouge.

Michelle, André, and Simone read during the ceremony. Patrick and John, our two sons-in-law, read as well.

We did not have James' body to bury, so we had honorary pall bearers led by Steven. Other pall bearers were: Brandon Allen, Doug Alongia, Brett Bordelon, Craig Clement, Adam Delaune, Luis Erazo, Chris Vizzini, and Rhett Ward. One of our nephews was an altar server. Bobby Helou, a dear friend of James, should have been asked to be an honorary pall bearer. I realize how close James and he were, and I have always wished I hadn't overlooked him in all the busy-ness of our preparation and grief. We talked about it at a meal some weeks later that we had for James' friends. I remember sitting with Bobby on the front porch and apologizing for the oversight. He was so kind and understanding … just one thing I wish I could have redone.

My sister, Paula, with her husband, Danny, provided a beautiful reception at their home after the memorial Mass. Many people gathered and shared delicious food and memories of James. I wish I remembered more about it, but I know it was special. And we were grateful.

***

Upon our arrival in Hawaii, an interim, prerequisite stop on the way to the RMI, we were greeted at our hotel by a Coast Guard representative, the same man with whom contact had already been made, Commander Mark Morin. This young man impressed us immediately with his warmth and compassion. In speaking with us, he was emotional at times. He explained to us that he was directly involved in the search for James and the others on the boat who were lost. He himself had a wife and three sons, and I remember him saying amidst the fog in my head that he couldn't imagine our loss. Jim and I, along with our children, met with him and others from the Coast Guard at one of their offices. They talked us through the search and were available to answer our questions. I honestly don't remember if we asked any questions or, if so, what they were. They also presented us with a binder they had compiled for us with the US Coast Guard 14th District seal and bold lettering on the front. It read, "OVERDUE SKIFF WITH 4 PERSONS ON BOARD IVO MAJURO, REPUBLIC OF MARSHALL ISLANDS." All I recall thinking was that I would have said there were five people on board since the young lady on the boat was pregnant. She had been on her way to a physician appointment on the main island. One page in the binder was a full-page color photo of the boat that was titled "Search Object—Picture of actual vessel after recovered by PPB." All of the pages were covered in plastic sleeves. Someone in that office cared enough to present us kindly with a summary of the search efforts to help us see and understand all that had gone into the search for all of our loved ones who were lost. I wonder who compiled that binder for us. I don't think I even asked. We were grateful.

We were in Hawaii from December 13th through the 15th to meet with the Coast Guard and to wait for an available flight to the Marshalls. Our family was able to get on that next flight to Majuro on December 16, 2010. We spent a lovely, warm evening in Majuro. Everyone was so kind. From there, our trip to and from Arno Atoll occurred on December 17th. I remember we barely made it back in time to board our flight out of the RMI. We were wet and pretty much a mess from our boat trip, but we boarded the plane—complete with wet papers, mementos, and a broken container holding it all, to head back home through Hawaii to Christmastime in Louisiana.

Commander Morin and his kind wife, Adele, whom we met years later in Virginia, stay in touch with us still. A more outstanding representative of our servicemen I don't think I could find. Over the years the Morins transferred with the Coast Guard as a family, from Hawaii to Alaska to Virginia. We have had the pleasure of watching Mark advance from commander to captain and Adele transition from teacher to nurse in the time we have known them. Their precious sons, Jack, Nick, and Max are fortunate young men to have these two amazing parents. Jim and I made a road trip a couple years ago to visit a number of friends in several states, very especially among them the Morins in Virginia. We went to their home for a lovely and delicious Italian meal and a visit that made us feel like we had been close friends for years. There we met their three sons and finally met Adele. I say *finally* for Adele because we knew her through mail, email, Facebook, and other kindnesses she had extended to us before we met her in person. What a gift the Morins have been and are to us.

***

Two memorial services were scheduled in the Marshalls, one on the main island in Majuro, the capital city, and one in Bikarej, the village on Arno Atoll where James was stationed and lived. Each, according to local customs, was captivating. I tried to pay attention as all who were lost on the boat were remembered in Majuro and as James and his *Baba* and all were again celebrated in Bikarej. On the main island, we were told that it was customary for women to cover up by wearing what I would describe as a muumuu. All of us females bought one. Mine was brown, and the weather for the service on the main island was intensely hot. I remember thinking that I was literally about to faint as I sat through that very long ceremony under the pelting sun. I truly don't think I ever felt that intensity of heat. In Bikarej, we celebrated in the little chapel of the village, and so we were inside (though with no air conditioning or power), and we were served coconuts there.

As a part of our travel to the Marshalls for these memorial celebrations, we were able: to see James' living quarters and envision better his life there; to meet many of his co-workers and families; to pick up some of his belongings, such as his few favorite clothing articles; to recoup letters and papers; and to gather some mementos. His kindle, phone, and camera were nowhere to be found. We assume they were with him on the boat, and so any last photos would never be seen (note: letter dated 08/20/10). And, as described in James' letters and noted previously, his snorkel booties for coral and lagoon exploration had been 'borrowed' early on. James let us know that, and he requested a new pair from us that we subsequently sent and he received. When we left, we gifted those new snorkel booties and a few other things to the families with whom James lived.

We were told before we went to Bikarej that the children (and adults) loved treats, especially candies … and especially chocolates. So we took bags of candies with us. We were told we would be able to throw them *Mardi Gras-style* in the chapel after the service. So we did. The children and many adults scrambled for them as people do for beads and doubloons at Mardi Gras parades. It was a taste of home in the RMI … and to me a demonstration of the universality of people.

In both locations, we were gifted with beautiful *amimono* by many, and we were even offered American dollars from a few people. In both places, there were baskets and boxes for collection of these crafts for the survivors of all who died, as is apparently customary. Small children brought gifts to us and smiled at us as they gently placed them in our box. In Bikarej, they loved James and we heard them utter his name *(Jam—iz)* in conversations we otherwise could not understand.

Upon leaving that second memorial service in James' village of Bikarej, the captain of our boat made a poor choice of maneuvers, sending us into a wave at a bad angle. Our boat, with all of James' siblings, Jim, me, other volunteers, and James' things that we had gathered to take home, began to take on water. We were just reaching the reef around the atoll, past which the water gets notably rougher. As water came in and James' belongings began floating, I remember Annie reaching frantically for James' things. Steven lost his hotel key and dove intentionally into the water to retrieve it. Simone panicked and froze. Was this really happening? Children from the shore began swimming toward us! Where were their parents? A feeling of total lack of control and general panic ensued. Some cursed; some prayed. I had a boot on my leg from a stress fracture, and I remember ripping it off so it wouldn't weigh me down.

Steven was back in the boat—with his keys in hand. James' papers were wet. A Marshallese man was talking in this language we couldn't understand and bucketing water out of the boat. Annie said something like she still couldn't believe this had happened once with James and it certainly wouldn't happen again. Then it all stopped.

Our boat was righted, water seemed to be mostly out of the boat, papers were sopping wet but mostly retrieved, our bin of *amimono* was still there, my boot was ruined, and everyone was okay. The ride the rest of the way to the main island was a bit bumpy—and utterly quiet.

I wonder why this occurrence happened to us. Were we supposed to feel what James felt? I remember thinking how additionally frightening it was to have all of our adult children there. I thought how scary it must also have been to James to have no one on his boat who spoke English when they began to lose control. Did he pray? Did he try to help … maybe try to help the pregnant woman? Did he suffer? I remember thinking that, with the intensity of the roughness of the water, the Pacific Ocean seemed inaccurately named. And one of our children said she hoped maybe that roughness let James get knocked out so that he didn't suffer. What was it like? Those words and images and questions hung in my head for a long time.

I wanted—felt like I needed—to allow my faith to help me stop thinking that way. Then a sweet friend in my Cracked Pots grief group, Trisha Brown, said to me one day something that helped me change my focus. She suggested that perhaps I could think of that boat—that awful image in my head of the photo of the ghosted boat after it was found with no person on it, bobbing in the Pacific—not as where James suffered and died, but as where James met Jesus. I am forever changed by the words of this dear friend, who grieves for her son the same way we grieve for ours.

***

After these celebrations of James' life, other tangible memorials began to happen. These have benefited people in the Marshalls, in Louisiana, and in Mozambique's Mission Amatongas. So much good has occurred in these places that I have devoted sections of this book to them in the chapters ahead that begin with the words "Of The Fruits of James' Life … " To those of you reading *Jambos*, I hope they inspire you and bring you peace.

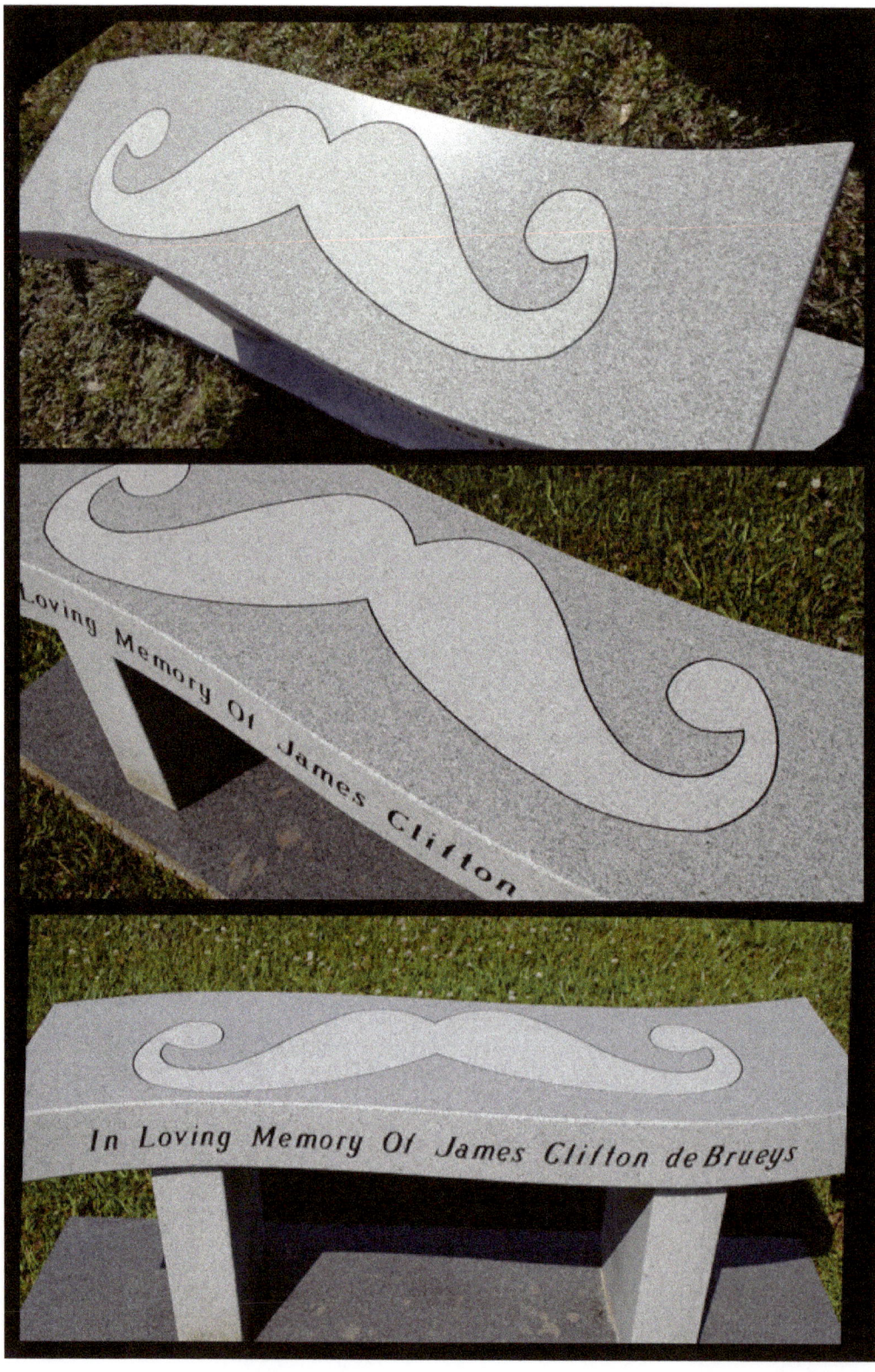

JAMBOS WITH JAMES

17

Of the League

While James was at LSU, he and three close friends: Doug Alongia, Luis Erazo, and Emily Eck, formed what they called The League (note: letter dated 11/02/10). The idea of the group melded into a particularly special yearly party with friends around Christmas time. The "classy" part of the party was a directive to dress up, wearing suits or even tuxedos or parts of tuxes. In the beginning, from 2007 through 2009 gatherings were at our condo. (This is where André and friends, later James, and then Steven with friends, lived near campus.) In 2010, immediately after James died, a small group of intimate friends gathered at Luis' for a League tribute. For as many parties as Jim and I have attended, we had never attended one with James, since we, as parents, only became a part of this event after his death. After 2010, the party has been hosted twice at Rachel and Doug's and twice at Brett's. Others were a year each at Jess', the Solinos', Adam's, Craig's family home, and Ali's. The location for 2020 remains to be determined.

After James' death and as some friends married and began families, the parties continued, in part in memory of James, while becoming inclusive of kids/families and other friends. Usually the children, along with us older folks, go early to visit and share a pot-luck meal. Some of James' siblings and our grands have gone with us pretty much every year since, though we missed 2019 (because of the pending birth of Simone and Jon's twins). We look forward to the 2020 event.

Enjoy the photos of the early and more recent Leagues! The tenth anniversary of the party was in 2017. I found this facebook entry from James the first year the League formed.

<u>James de Brueys</u>
20 September 2007- As the weather cools down, the time for a gathering draws nearer... gentlemen and ladies ... suit up time is nearing like our inevitable doom.

<u>Luis Erazo</u>
6 July 2007 - James, you have the right spirit. I'm off to search.

James de Brueys
6 July 2007 - Dude, Goodwill in BR has some choice suits. I got a 3-piecer... for 10 bucks. Suit search. No matter the type. NO MATTER WHAT IT TAKES

18

Of Visitors from World Teach Who Traveled to See Us in Louisiana

In the late fall of 2011, our first visitor from James' World Teach group, Clare Jones, visited us with Miranda Powell and her children. Clare and her family lived in New Orleans. We learned then that she was the only other Louisiana volunteer in James' group who had flown in July 2010 to California to join the others traveling to the RMI. James and Clare had never met before, but Clare told us later how at ease James made her feel as they introduced themselves and learned they had the same destination. In May 2014, Clare dedicated her graduate school thesis for the University of Iowa, *Wager*, to James with the words: For James deBrueys.

Lara Farina was the next visitor to come to Louisiana from James' volunteer group. That visit was in 2012, shortly after the one-year mark of the deaths of James and the others on the boat with him. We picked her up in New Orleans and got to have her stay with us in Baton Rouge. Lara and James seemed to have had a special relationship. At the airport meeting room where we first met many of the volunteers in Majuro, I remember noticing her sadness in particular. I can see it in one of our photos taken there, too.

On the heels of that visit, Brooke Payne visited in February 2012. Her visit coincided with the dedication of our memorial garden at St. Aloysius Catholic School. We were so happy to have one of James' fellow volunteers with us to help turn the soil to mark that garden's dedication.

In April that year, Mandy Doyle came to New Orleans and we met her there for a classic po-boy meal in Uptown NOLA.

August 2012 marked a visit from Erica Moore and her boyfriend. Our entire family had a memorable lunch with them in New Orleans near City Park.

We began 2013 with a visit (the first of two) from Laura Sundblad. She came to Baton Rouge and stayed with us. We visited a haunted plantation restaurant outside Baton Rouge during her stay.

In December of that same year, Miranda Powell and her children came to celebrate with us at our Open Home in Baton Rouge.

In the summer of 2014, we had a special visit from Julie Walsh. She was born and grew up in New Orleans and lived currently in Hawaii with her two sons. How her story intertwines with ours comes later. That was the first time we met her, but it wasn't to be our last visit.

Deborah Hallen and Paul Zelinsky visited Jim and me in New Orleans in January 2015. We had a lovely meal and got to know them as parents of Anna, a previous volunteer in the Marshalls.

In March of that same year, Laura Sundblad came for a second visit. That she made another visit and continues to make an effort to stay in touch with us is incredibly special.

In the summer of 2015, Simone and I made a trip to New Orleans to see Miranda. Julie Walsh was in town, too—what a bonus! We went to see about the Marshallese wedding flowers for Simone and Jon's wedding, which Mona and David Strauss were gifting to them. The *amimono* flowers for the groomsmen, the bridesmaids, and very especially and beautifully, for Simone's wedding bouquet, created through the artistry of the ladies of the Marshall Islands, were remarkable. Simone's bouquet was complete with a small medallion Mona had made, which had James' picture hanging delicately from it. (You may recall that Mona and David were there for us in Majuro when we arrived at the airport, throughout our stay, and when we departed to return home from the Marshalls.)

Miranda Powell came to visit again with her sister, Pele, along with Katie Finberg Fotofili, for our Open Home in December 2015. That was our five-year mark, so visitors meant so much.

In December 2015, we missed a visit with Anna Zelinsky and also missed what would have been a second visit in January 2016 with her parents, Deborah and Paul. I had bilateral knee replacement surgery around that time and wasn't mobile. We were sorry to miss these two opportunities. I was in the hospital on James' birthday and through Christmas at this five-year mark, just as I had been with James the day of his birth and through Christmas that year. This surgery date and hospital visit somehow felt right.

In the summer of 2016, we were invited to Julie Walsh's parents' home to visit with them, Julie, and her family while they were in from Hawaii.

In January 2017, we had a lovely gathering at Miranda's home with her family and Mona and David, her parents.

Then, in June that same year, we welcomed the amazing Ken Hagberg to New Orleans for a meal and visit. We even got to meet his parents and siblings later in the day for coffee and beignets.

In September 2017, Mona and David came in again from the RMI, and we met for seafood on Bayou Manchac between Baton Rouge and New Orleans.

September 2017 saw Katie Finberg Fotofili return to New Orleans for a conference, so we got to intercept her for lunch in the city on her break … that I believe we extended a bit beyond her allotted lunch break.

Clare Jones and her dad came again to see us in Baton Rouge for our Open Home on November 25, 2018.

Life goes on. We expect that visitors and contacts will wane over time. These visits have meant so very much to us. We remain grateful for calls, emails, mail, and Facebook that keep us connected always. We are excitedly anticipating, after canceling our 2019 Open Home due to the twins' arrival, that we will have special visitors in our home again as we celebrate James during our Open Home at ten years on November 29, 2020! The effect of covid remains to be seen by that time this fall as this book is published for James' 10th anniversary. Who knows …we are thinking we may have a drive-by party with food and party treats to go. James would smile on that!

19

Of James' Namesakes

We can never say what might have been. Would James have married? Had children? What we do know is that he lives on in us and others. Since James' death, several family and dear friends have named their children for him in his memory. We are so appreciative.

Samantha James McCrary was born December 3, 2012. Sam is third of our now six grandchildren. Desiree Costillo, a special friend of James, told me shortly after James died that she saw him in Heaven. She was ill and had coded on an operating table in a hospital in New Orleans. During that time, she said she saw James and that he gave her three special messages. One of them was that, whenever we are with Samantha James, he is there with us very especially. A second of Desiree's messages from James was personal to her family. The third one was to tell me he was okay and to not cry too much.

Two of James' closest friends gave their newborn boys James as their middle names. James, throughout college and beyond, had participated in many antics and rites of passage with two close friends, Rachel and Douglass Alongia. He even sang informally at their wedding. Their second child was born Dougie James Alongia on December 22, 2013. Two other special high school and college friends, Allyce Baudier and Brady Lemon, named their first child Xander James on January 22, 2014.

Before James had left for the Marshalls, he was honored to be godfather to Marley Bordelon, who was born on February 1, 2009 to Brett Bordelon and Ali Hymel. Brett was an amazing friend to James from the time they were little boys. He and Chris Vizzini, the three amigos from Monte Carlo Drive who were pretty inseparable, are referenced throughout this book.

Then there is, last to be mentioned but certainly not least, the final "little James." Imagine the gratitude and honor we felt when we learned about little James when he was already almost five years old. He was the first baby born in Arno after James died, and his family named him James (pronounced *Jam—iz*) by his family, just as our James' name was pronounced by Marshallese children and parents during his months in the RMI.

JAMBOS WITH JAMES

20

Of the Fruits of James' Life …
In the Marshall Islands (RMI)

The full basketball court that James envisioned and hoped to provide for the Bikarej community became a reality (note: letter dated 10/05/10). Money started coming to our address via James' suggestion, and family and friends stepped forward to help in incredible ways. Most outstanding was the work of Ken Hagberg, one of James' peers in his World Teach volunteer group. Ken stayed beyond his assigned months in the RMI to begin and finish the court. He was a cheerleader, taskmaster, time manager, and engineer. From grinding stumps in the jungle to making concrete with sand, to organizing labor among the families there … a lot of blood (literally), sweat, and no doubt tears (when the task looked impossible) got it done. Not only did Ken put a hundred percent of his energy and enthusiasm into this endeavor, he sacrificed a bit of his own well-being to see it through. When Ken returned to Illinois, his home state, he was hospitalized for infections in his feet from exposure to the elements in and around the jungle during the building of the court. Since the completion of the court, our family had the honor of a visit from Ken in Louisiana. As a bonus, his family was traveling through New Orleans, so we met his parents and siblings as well. The court began in December 2010 and was completed early spring 2011. That's pretty amazing under any circumstances, I think, and especially within Marshallese time!

James' brother, Steven, made a second trip to the Marshalls to dedicate the court in memory of James and all who had died on the boat with him. Very dear friends of ours, Kim and Bobby Shackleton, donated the memorial sign, made of Louisiana cypress, and shipped it to the Marshalls for the dedication. Jim and I designed it with Dwayne Broussard of Patterson, Louisiana, who then crafted it beautifully. Lynn and Denny Otillio very generously donated monetarily to these efforts. Jim and I cannot adequately express our gratitude to our amazing and loyal friends.

There have been opportunities for contributions to other efforts in the Marshalls. One only needs to Google such opportunities to find want for hospital additions, materials and books for schools, refurbishing of schools, amenities such as our basketball court, and a myriad of other ways to help in this and other remote parts of the world. Helen Claire sent me a copy of my letter sent to her and Lisa, who was in charge of World Teach bookkeeping, in January 2011. It read, *Annie and Angela talked with us while we were in the Marshalls about James' host family who lost their dad. We would like, if possible, to send $30.00/month for 12 months to them (which turned into several years) specifically to help them with food and other expenses. If funds could be earmarked for that purpose from the friends in Louisiana who wanted to help, the remaining funds we would like to have go to the Marshalls/Bikarej for our projects as needed. Many thanks!*

Because of James' death, World Teach began requiring volunteers on boats to have life jackets with them. We were told that the vests provided were outfitted with equipment to detect their presence in the ocean. Our family was pleased that awareness efforts had increased and that life jackets were provided. But this is not enough. Jackets should be required to be worn, not merely to be on the boat. Evidence of this need was pointedly demonstrated during our family's venture as mentioned earlier when we traveled from Bikarej and then returned to Majuro on the day we left the Marshall Islands.

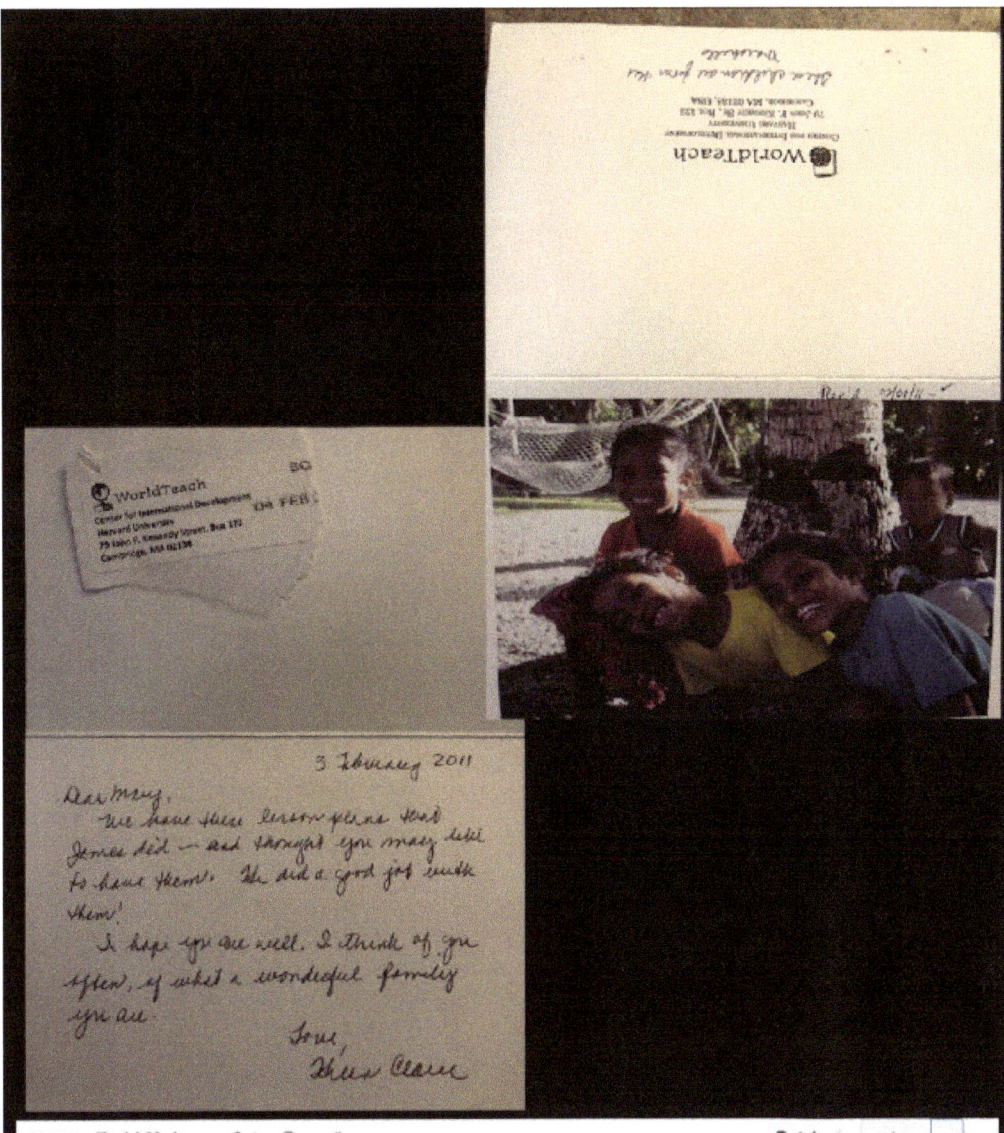

Todd Mulroy <y2ytpp@gmail.com>
to marydeb, bmcspadden, WorldTeach

Hi Erin,
 The fund was used to build and upkeep the James Clifton DeBrueys Basketball court in Bikarej, Arno. The funding started in 2011 and has been on going ever since. The the basketball court is now finished and has recently been touched up with money from the fund. Over the years it was asked of me by Mary T DeBrueys (James Mother) to send food and supplies to James Host Mother in Bikarej.
 I am not sure how much is left in the fund, but I have included Mary T Debrueys in this email to help us figure out what the remaining funds should be used for.

Hi Mary T,
 I hope all is well. I am no longer with WT and Erin Girard has replaced me as Field Director. Would you like Erin to continue to provide food for the Joreum Family or would you like to use the funds in a different way? I am willing to assist Erin in anyway possible to see the funds are put to good use. Please advise when you have a moment.
 Please tell everyone I said hello and that I have returned to the RMI after a quick trip home and am currently working at Youth to Youth in Health. All the best to you and the DeBrueys Family.

MARY T HEFFRON DEBRUEYS

Lesson Plan 1 (Oct 9 & 10) — James deBrueys — B. Karej

Objective: SWBAT ① identify a play based on its structure, ② define the words "play," "act," "scene," "actor," "actress," "narrator" by completing a worksheet. ③ Act out a play with proper intonation.

Material — dictionary, textbook (Shining Sun)

Step 1: Explain a play (after assessing their own knowledge) is a "movie performed live." Describe its parts in general and write the terms on the board.

Step 2: Groups with dictionaries look up the terms. Define them on the board, explain further. Look at textbook with play (Jewel in the Sand). Explain the roles they will take on and act out a few parts myself, have them read it through.

Step 3: They perform (book in hand), knowing which part belongs to them. Switch actors and perform again.

Step 4: Explain how many plays are longer and address our new terms asking for an explanation.

Step 5: Final assessment = ① acting out play, ② matching definitions on a key.

Student interest will be maintained through looking up definitions in the dictionary as teams and performing the play as a class.

[#1]

[Oct 18-22] — [Grade K-1] — James deBrueys — B. Karej

Objective: SWBAT ① introduce themselves ② sing the alphabet (new song)

Materials: alphabet chart

(Side note: the second quarter will start my new English class with grade K-1. I will be assessing what English they know and will be dealing with the inevitable behavior problems.)

Step 1: Introduce myself to the class, slowly and with proper pronunciation. Ask them what their names are. If they do not follow, I will ask the teacher present in the room and model an introduction. For the alphabet, I will use the chart for a visible model and ask the teacher to help me model the "new" alphabet song (rock & roll style).

Step 2: Have them write/draw their names. Model my name and show my drawings (have the teacher do it too). For the song, maintain practice and associate letters with the song "lyrics."

Step 3: Have them practice introducing themselves to their neighbor (model with students prior). Point to letters and have them verbally identify the letter.

Step 4: They will each introduce themselves to me (or I to them). They will sing the song without my aid in a "performance" (possibly during recess for the other students).

[#2]

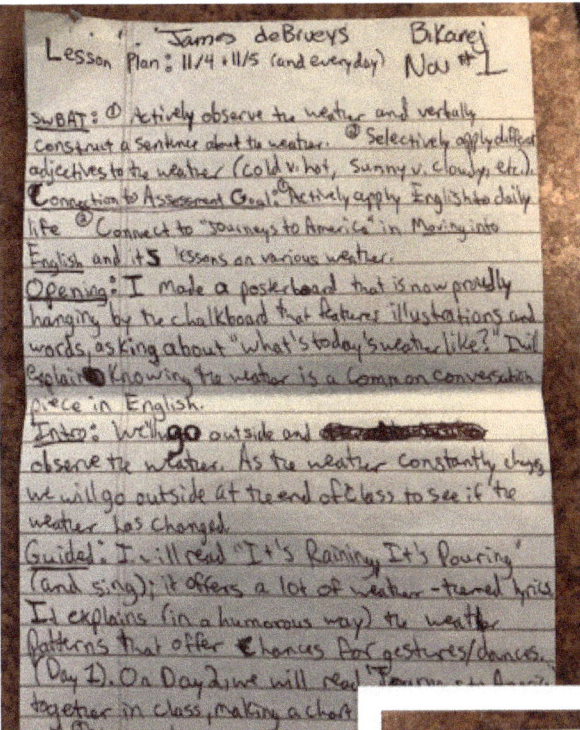

MARY T HEFFRON DEBRUEYS

11/1 – 11/31 – Letter Recognition/Sound Recognition

SWBAT: ① orally identify all the letters of the alphabet in random order ② orally make the sounds of the letters (ex: 'b' makes "buh", 'p' makes "puh") in random order

Assessment goal – have students recognize letters and sounds. This meets the expectations set by the MoE and prepares them for reading (both English and Marshallese).

Open: As I only have 30 minutes twice a week to work with grade K-1 and 2-3, and limited attention spans, this piece takes about 10 minutes per class. I put letters on the board (2-3) and ask "what letter is this?" I then explain the letters also make sounds.

Together: The students stand as we "perform" an action associated with the letter. For example, 'b' is like "ball" we pretend to shoot a basket. "B!" Then I say "what sound does it make?" and together, while shooting the basket, say "Buh!" We go through the letters, and repeat multiple times. We practice singing the alphabet with the gestures.

Independent: I have the students sing the alphabet together, and I listen for pronunciation and look for the proper "moveout" with the letter. Then, I have made 52 "alphabet cards" of capital and lowercase letters. I can ask "what letter" or "what sound" and look for proper gestures and (more importantly) the right letters.

#5

Assessment: I anticipate (next month) to practice properly writing the letters. Assessment will come, currently, from having students simply recognize the letters. (My ultimate goal is to have them write the letters when I orally call them out).

Conclusion: When we complete the alphabet, I will congratulate them for their great pronunciation. I will tell them we will practice more reading now that we definitely know the alphabet!"

Materials: ○ 52 sheets of construction paper with one letter (capital or lowercase) on each one
○ alphabet displayed on the wall
○ dance moves

✱ Thank you for the help on this lesson plan, Angela. It has been working out.

[Grade K-1 and 2-3.]

— James de Brueys

#6

21

Of the Fruits of James' Life …
In Mozambique

James and Steven, while high school students at Brother Martin, were taught by lay teachers and by the Brothers of the Sacred Heart. One of their favorite brothers in New Orleans was Brother Christopher Sweeney, a young, athletic brother who used to shoot hoops with the kids after school.

A few years after James' death, Jim and I reunited with Brother Chris. We didn't know him well yet, but we felt compelled to assist with the efforts of the brothers at their mission in Africa. Together with other brothers of the same order around the world, they built and maintain Amatongas Mission in Mozambique. The children served there are among the world's poorest of the poor. Some are orphaned and some are day students.

Jim and I began a fund through the Brothers of the Sacred Heart Foundation to support Brother Chris and the efforts of his fellow brothers in memory of James. The JAMSS fund was begun—standing for James' Amatongas Mission School Supplies. (Years ago as a family, we named our first dog Jamss, a name composed of the first letters of our five children's names in scrambled order—James, André, Michelle, Steven, and Simone.) Donations made to the JAMSS fund go to Amatongas for school supplies and any needs directly related to the students and boarders at this orphanage and school. Among other uses, a priority for these funds is the purchase of malaria medication. We understand that this medication is in short supply and can often be over-priced due to the shortage. Children can be sponsored with monthly support donated to Amatongas by individuals as well.

We began our support of Amatongas in 2013, and James' fund began within the year after that. Following is a letter we have had occasion to send to those interested. It is an invitation we like to extend:

The opportunity for a Memorial contribution to Mission Amatongas in Mozambique in memory of James is a reality!

Anyone who would like to donate to this orphanage and school run by the Brothers of the Sacred Heart in Africa in James's memory can do so by sending a check to:

Brothers of the Sacred Heart Foundation
4401 Elysian Fields Avenue
New Orleans, LA 70122

Please consider making a contribution! Any amount is so helpful, and 100% of what you send goes straight to Amatongas. If you would like to help, please write your check to:

Brothers of the Sacred Heart Foundation.

On your check's message line, please note: "JAMSS" ("James' Amatongas Mission School Supplies").

Brother Chris Sweeney in Mozambique will use these donations for school supplies and as needed. He is free to use the funds for pressing needs for the children at his discretion. This could include: malaria medication, clothing, food, seeds, water, etc.

You can read more about Amatongas @http://amatongasbound.blogspot.com/ or look them up on Facebook @ Mission Amatongas.

The precious children at Amatongas are truly among the poorest of the poor. Their orphanage and school now serves girls, which Jim and I feel is a huge step forward in this culture and so important!

There is much need there; please consider helping.

A contact number for the Brothers of the Sacred Heart office is 504-301-4758. My office # is 225-388-9822. Please call either one of us with any questions.

And PLEASE feel free to share!

Many, many thanks!

Mary T & Jim

22

Of the Fruits of James' Life …
In Louisiana

An anonymous donor within St. Aloysius Parish, our current church parish in Baton Rouge where I attended elementary school in the 1960s, offered to sponsor a memorial scholarship for a teacher each year. He or she is chosen as one who "thinks outside the box" as James did when he taught. Each year one teacher receives a stipend to be used half as a personal gift and half to inspire and provide something special for her students and the school community.

The memorial garden at St. Aloysius was begun by the first teacher who won James' award. This garden produces flowers, vegetables, and fruit—some of which are consumed by the students in the cafeteria's lunches. We have had a couple dedications within the garden, such as the one attended by family and friends shown in our photo collage.

Winners since the award began, beginning with the 2011-2012 school year and through the 2019-2020 academic year are: Carolyn Murphy (who started the memorial garden), Jane St. Pierre, Michaelyn Bellelo, Lisa Mongrue, Joanna Lemoine, Cheryl Frost, Karen Edwards, Stephanie Coxe (whose students self-published creative books dedicated in James' memory and shared them with a local school), and Tracey Barhorst (who, with her students in 2020, started a café in memory of James to be set up on the day her sixth graders honor the French culture). After the publication of Mrs. Coxe's class's books, a neighbor's son, Adam, presented us with one for which he was the artist. We are grateful to St. Aloysius School for all they do to honor James' memory in such remarkable ways.

Also in Baton Rouge at St. Aloysius, our church sponsors a non-denominational women's group begun for mothers who have lost adult children. This group, named *Cracked Pots … Blessed & Broken*, began with five members under the guidance of Lisette Borné. She led us for some years before yielding the leadership to me so that she could pursue other diocesan responsibilities. Though she was the founder, she was the only member of our group who, in fact, had never lost a child. In spite of that, she felt called to start the group when several of us lost children very close together in time. Our group now has over forty members. *Kintsukuroi* is a Japanese art form we adopted to represent our group. It is the art of filling cracks of broken vessels with gold to make them stronger. This beautiful art represents to us that vessels can become more beautiful for having been broken when filled with the gold of our faith.

***

Positive outcomes have been appreciated in the Kenner/New Orleans area of Louisiana. St. Elizabeth Ann Seton, our pre-Katrina church parish, and the elementary school James, Steven, and Simone attended in Kenner, is the location of a beautiful memorial dedicated by a group of parents, including us, who have joined together in this club no one ever hopes to join … of parents who have lost children. Our memorial is dedicated to those who have died who attended the school and/or participated in the parish. The support and the relationships within such a group are priceless gifts on this journey.

Each year at his high school, James is remembered with many others on the St. Joseph's Altar at Brother Martin. Thanks to Cissy Yakelis who keeps James' framed photo for us at the school and ensures that it is put back on the altar for prayer each and every year.

***

Shortly after James' death, the Marshallese children at one of the schools participated in a writing exercise. They wrote and published fourteen books in total, one of which was called *Taki and His Boat* in memory of James. That specific book teaches the importance of water safety. In the spring of 2013, our family donated three sets of this book series. On our family's behalf, Steven delivered a set of these books each to the libraries of Brother Martin High School in New Orleans, St. Elizabeth Ann Seton in Kenner, and St. Aloysius in Baton Rouge.

JAMBOS WITH JAMES

23

Of Life Going On & Arrival at the 10-Year Mark

From throughout the early years, middle and high school, college, and since James' death, precious friends remain.

Some family members and some friends of James, who have become our young friends as well, have sent us some of their fondest and most outstanding memories of James to help me represent them in *Jambos*. Here are a few in no particular order:

I first met James in high school, going to Brother Martin and Chapelle dances with groups of friends. We had a lot of mutual friends, especially Craig Clement who I went to grammar school with! I remember hanging out with James and Craig at LSU Spring Testing. I was stressed out and they were just hanging out in the union, being goofy as usual! I remember being glad I ran into them! David and I started dating at the beginning of freshman year at LSU and he was friends with James from Brother Martin. We would spend a lot of time with James and the other Brother Martin guys! My favorite memory is going to Mellow Mushroom one night to listen to James play with his band, and James singing "Play that funky music David Gambel." So funny!! I also ended up taking a Latin ballroom dance class with James at LSU! It was either junior or senior year. He had some great moves! We love James! I cannot believe it will be 10 years. I can only imagine how emotional this Jambos project must be. What an amazing way to honor James!

- Kim Gambel (2020)

You and James are always in my thoughts this time of year.

There were so many nerves our first few weeks on island, during orientation, and James was so open, friendly and genuinely interested in getting to know me. I really appreciated that. I was lucky to know him, even for just a short time!

Thanks for keeping in touch all these years. Joe and I still need to plan a visit to NOLA!

- Katie L. Fotofili (2019)

You would not believe the eerie squalls we have had here yesterday and today (Thanksgiving here). Exactly the same as 9 years ago.

- David Strauss (2019)

I think my favorite story of me and James was one time we hung out after he got off work. He took me to Pinetta's to have some wine. He told me that Pinetta's was a great place to take a date. I remember when we walked in or maybe it was when we were leaving somebody yelled "Amish" at us because of our facial hair. Haha.

- Sean Lou (2019)

I met James 13 years ago when we were just freshmen at LSU. I practically lived with him for a whole year, playing video games, listening to his cover band play Red Hot Chili Peppers songs, and got to know truly one of the most amazing human beings I ever had the pleasure of knowing. I remember I left for a while, when I came back to Baton Rouge I called him up to catch up on old times ... he was at work at the radio station and that day he played a song just for me... I'll never forget listening to the radio in my car, "This one's for you Sara Sisto."

Anyone who ever knew James or got the chance to meet him for just one minute knows how special he was; how immensely smart, kind, hilarious, warm, and generous he was. I think about you a lot James.

It was around this time I also met Kaylee; another beautiful soul that left too soon. I remember Kaylee fondly. She also had a way of making people smile and laugh. I pray that James' and Kaylee's souls have reunited.

- Sara Sisto (2018)

They (James and Kaylee) were the first couple that were actually able to understand what unconditional love was from my age bracket. At 15, I admired this. James and I would be at parties, and he would tell me how cool it was to have your best friend be your girlfriend. I laughed at him but damn, how right he was. When James passed, I didn't reach out to Kaylee for a year out of pure and utter pain. Didn't know what to say. Didn't know how to say it. So when I saw her, finally, she hugged me. And I cried. And cried. And cried some more. Now, I find out you are gone as well.
And I am sad.
But, somehow, it's not your ordinary death.

Cause my boy has his angel back with him.
And that's worth smiling over.
RIP Kaylee

- Scott Patrick Hogan (2018)

My brother, James, disappeared into the Pacific 7 years ago today. It wasn't all for nothing. After our family visited the islands, so many wonderful human connections were made. His selfless soul still inspires.

- André deBrueys Cardinale (2017)

My first strong memory of James is driving with him and Steven to Brother Martin High School, not sure when or why, maybe morning to class or to a football game? But he introduced me to two bands I had never heard of: Interpol and Cake. They are both now personal favorites. Makes us think of him when they play.

- James Catalano (2020)

I met James working at KLSU. He had such a good energy on air and in person that everyone who met him just fell in love with him. He was amazing as a DJ and was a friend to everyone. James was one of those people that was like the fiber that was woven straight through the tapestry of the radio crew. He was friends with all the people and many of the girls really liked him, a couple even dated him. James was smart, funny, handsome, and talented. We were lucky to have him with us in the studio for the time we all got to spend together. What great memories! He was so good on air!! I worked there 2008-2010.

- Grace Doll (2020)

This popped up in my facebook feed from 7 YEARS AGO TODAY! I love it! James was at LSU.

James de Brueys to Mary T Heffron DeBrueys
4 March 2010 at 19:01
well hello mother

- Mary T deBrueys (2017)

I worked with James at the Chimes. I was new, and people that worked there weren't always nice to the trainees. James was behind the bar and so so nice and upbeat and happy with everyone, and was once of the only bartenders that was patient and kind with newbie mistakes. We were just coworkers but he made whoever he was talking to feel good. And he made you laugh and laugh he was just so damn funny. You don't often meet people that have a certain light around them but he had it. I've always thought the world lost a really good one when he died.

- Jillian Crawford (2020)

I thought of you today, but that is nothing new.
I thought about you yesterday and days before that too …
Your memory is a keepsake from which I'll never part.
God has you in His arms. I have you in my heart. (idlehearts.com)

- Desiree Costillo (2016)

I'll never forget what James told me and he's not the only one to ever tell me this but he's the only one I ever believed bc well it was James. He said, "You're worth it all, just know that, I don't lie." A lot of my life I blocked out but that's one thing that stuck with me—his opinion always mattered to me even when I didn't act like it.

- Brittany Collins (2016)

THE BOXES OF SUPPLIES ARE FINALLY STARTING TO ARRIVE IN THE MARSHALL SLANDS!! I just got word today that the first two boxes have arrived at Third Island. 10 more are already on their way, and I have enough money from donations to send at least 27 more!! Thank you for helping me make this possible!! To everyone who donated time to collect, wash, organize, pack, promote, and publicize, as well as those who collectively donated over $1,400 to pay for shipping costs, THANK YOU SO MUCH! — with Laura Atilano, Connor McCutcheon, Mary T Heffron deBrueys and 26 others at Kwajalein Atoll, Marshall Islands.

- Stephanie McCutcheon (2012)

Hello everyone (Iokwe noan aolep). I'm writing to co-workers and families in the Marshall Islands at the time of the two year anniversary of James' death. Two years ago a very special man, James, left this world, but he left it better than he found it, especially for his family and the kiddos in the Marshall Islands. (Ruo ila remootlok juon emman elap an joij , ekar jen lal in nan ippan Anij, ak ekaar komman bwe lal in emanlok jen mokta, elaptata nan bamili eo an kab ajri ro ilo Majel.) Please keep James and his family in your thoughts and prayers, learning from James and striving to do the best we can to make the world a good place. (Jouj im lomnak im jar wot con James im bamili eo an, kwon mour ainwot James ekaar mour, im command bwe lal in ej joun jikin eman.) We love you, James, now and forever! (Kimij lokwe iok James, kiio nan indeeo!)

- Kristina Bramwell (2012)

Hi Mary T and Jim (and family)!

Thank you for the birthday wishes—receiving them from you every year is something that means very much to me and I look forward to it every year. This year I was fortunate and blessed to be able to spend the day on a small island in the Chuuk lagoon reflecting and appreciating everyone important in my life and I certainly thought a lot about you and James. Please let me know if there is anything you may need from me or any way I may be able to help as you continue working toward the writing project.

I hope all is well in Louisiana and I'm looking forward to hopefully making my way back to see you in the next few years.

Thank you again and talk to you soon—Love Ken

- Kenneth Hagberg (2019)

We appreciate your note and kind words, Ken!

There is actually something you may be able to do to help me with the Jambos book. There is a little boy in the RMI named James who, we are told, was the first or one of the first babies to be born (on Arno Atoll?) after James died. He is named after our James!

We have indirectly kept in touch with James/his family through Todd and a little through Angela in the past. We have sent money, for example, to Todd to get James a Christmas treat each year. Since Todd is gone and WT seems to have taken a break or is now shut down (?), we have no one to help us stay in touch with James. Todd and Angela each said they would try to help, but we have gotten nothing back re: contact for this year. I know 'mail' doesn't exactly go to some of these families, but it's hard to accept that we now have nothing at all in terms of contact with this little boy! I want to have something in our book about him.

Also, I am thinking down the road a bit ... such as if he ever comes to the U.S. for schooling. We would want to know he is here, if we could help, etc. We have a couple pics of him and his Dad (attached). Is it possible you could help us locate him and find out if we could send him mail somewhere that doesn't depend on a 'middle-man'? I understand they either live on or go to the main island from time to time. Do you have contacts there?

If this is not possible ... no harm done! But if you have a way to help, that would be amazing! This has been a frustration for me for the last couple years. I am sad to feel we have no resolve.

Have a Remarkable 2020!

Love,

Mary T & Jim

- Mary T Heffron deBrueys (2019)

Hi Mary T,

I remember him (Little James) very well! It is nothing short of amazing that you have kept in contact for so long in a place where it's almost impossible, such a blessing. I've sent around messages to as many people as I could think of and will let you know as soon as I hear anything helpful back. In the meantime though, I heard that Todd is actually returning to Majuro from Arkansas this month. I wonder if he may be able to help us establish some sort of direct communication with him and his parents—set up a Facebook account or something similar. While I know Todd is returning, I'm not sure if he plans to stay long term or not so this may be our best opportunity. Todd could reach Bikarej through radios from

Majuro or he could to stop by the AM radio station and put a short radio announcement out over V7AB asking for the family or Bikarej leaders to contact him directly on a local phone number. I'll keep thinking about it and contacting people, I'm confident we'll get back in contact with them soon!

Happy New Year!

Ken

- Kenneth Hagberg (2020)

*Hi, Ken! Anything at all you can do, including if you speak with Todd, will be SO appreciated! I love the idea of Facebook or something set up directly so we can be in touch with the family instead of searching for middlemen! I just didn't know where to go from here and just wish I had asked your help sooner! By the way ... your pronoun 'we' (I'll keep thinking about it and contacting people, I'm confident **we**'ll get back in contact with them soon!) is comforting (!). Thanks very much for joining us in this effort!*

Love & Happy 2020!

Mary T

- Mary T Heffron deBrueys (2020)

Dear Mr. Jim & Mrs. Mary T,
Thank you for joining us to help make our wedding perfect. It meant the world to me to have you both there (my second parents), as we both know I wished I could've had James up there standing beside me. Thank you for always being such a positive force and inspiration in our lives. Jess and I hope to have such a happy, loving, and fulfilled marriage as you two.

Love, Rhett & Jess

- Rhett Ward (2020)

When Kaylee, James, Doug, and I went camping, one night we found ourselves deep out in the seldom traversed deserts of New Mexico. There were rocks, sparse cacti and brush, healed up cow bites in the cacti, and even a dead cow that had been left behind. It was just open desert and nothing else for as far as the eye could see. No buildings, no fences. The stars were clear and bright at night. We found a pile of rocks half-fashioned into a cooking area so we finished the job. We had turkey gumbo in the cooler Kaylee's dad had made for us and had brought all the way from home, vacuum sealed with love. We toasted tortillas on flat rocks. We had crammed things into every nook and cranny of that little car and the marshmallows ended up fused into one giant mallow, so we found a very big stick! It was very cold once we put the fire out that night and it felt good to be in the tent. Unfortunately, the fire flickered back up. What with it being a fire hazard season, and Doug and James being men of action, they ran out in their long socks and without hesitation ... peed on the fire!

- Rachel Alongia (2020)

I was going through my "treasure box" where I keep special items, keepsakes, and letters. I came across this beautiful celebration of life for James and I thought you may want another copy. It portrays his personality and love for people very well. It gives me peace knowing he had a family in the Marshall Islands and felt love from many during his time there. He was doing such noble work teaching, cleaning, and building the school there and his legacy will live on forever, especially when the church sings "Joy to the World." I know you are proud of James and we are, too … thinking of him often … Rachel

- Rachel Grammer Peterson (2018)

I remember his house mom in Spain always called him "chico con la barba." My 22nd birthday happened while we were there and I remember James took me to get an Irish coffee. This picture (James and Andrew on the ground) is my favorite picture of the 2 of us when we were in Spain.

- Andrew Herpich (2020)

Hey man I was driving home from work this morning and listening to Green Day's latest album. Felt like you were sitting right next to me singing the lyrics, just like when we were listening to that leaked copy of American Idiot months before it was released. I don't think that left either of our cars for months. Haven't felt your presence that strong in a long time. Miss you brother more than words can describe.

- Brett Bordelon (2020)

Love you man! I have these moments too.

- Brandon Allen (2020)

JAMBOS WITH JAMES

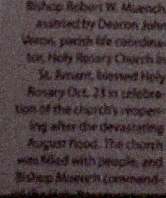

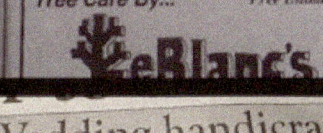

JAMBOS WITH JAMES

24

Of Narratives of Faith

> The souls of the righteous are in the hand of God,
> and no torment shall touch them.
> They seemed, in the view of the foolish, to be dead;
> and their passing away was thought an affliction
> and their going forth from us, utter destruction.
> But they are in peace.
> For if to others, indeed, they seem punished,
> yet is their hope full of immortality;
> Chastised a little, they shall be greatly blessed,
> because God tried them
> and found them worthy of himself.
> As gold in the furnace, he proved them,
> and as sacrificial offerings he took them to himself.
>
> Wisdom 3: 1-6

I have always been drawn to this reading at funerals, pretty much the only place I hear it. To me, it is down-to-earth and filled with hope.

***

Another beautiful writing is this one that our oldest daughter, Michelle, frequently quotes; I have come to love it. It gives me comfort.

> "The sole meaning of life is to serve humanity by contributing to the establishment of the Kingdom of God. *Find us ready, Lord, not standing still. Find us working, loving, and doing Your will.*"

***

Amazing, unbelievable, inspiring, anxiety-producing, questionable, faith-kindling phenomena happen to all of us throughout our lives. It can be difficult, in fact a huge leap, to see them through the eyes of faith. Some stories are private and don't feel meant to be shared. Some seem to be meant to be shared with only those close to us. Some seem meant for a greater good, to help all of us on our faith journey. Several such occurrences have offered me expanding hope since James died, and I hope they may do the same for you. I would like to share a few here.

This first one connects Spain with Louisiana. Shortly after James died, a friend of our daughter, André, knocked on her door in Madrid. She presented André with a ribbon from the Cathedral of Pilar. The Cathedral-Basilica of Our Lady of the Pillar (in Spanish: *Catedral-Basílica de Nuestra Señora del Pilar*) is a Roman Catholic cathedral in the city of Zaragoza, Aragon (Spain). The Basilica venerates the Blessed Virgin Mary, under her title Our Lady of the Pillar (*Pilar*). Visitors to the cathedral can obtain ribbons as keepsakes of prayer there that represent petitions and thanksgiving. This friend brought André a ribbon in memory of James. When her friend departed, André emailed me to tell me about her friend's visit and about Our Lady of Pilar. I had never before heard of this particular reference and title for our Blessed Mother. As I was reading her email, my phone rang. It was one of my coordinators from work who is also a dear and faith-filled friend. Carolyn Haar was calling to ask me if I had ever heard of Our Lady of Pilar and to tell me that she was feeling called to ask the Blessed Mother's intercession for James and for us. Sitting in my office in Baton Rouge, within minutes I heard this reference to Mary from two people—one in New Orleans and one in Madrid.

A short time later, Jim and I would make a trip to Madrid to visit André and John while they were still teaching English there. We traveled with them to the cathedral in Zaragoza. It is beautiful and it felt right that we were there.

***

Hola! Buenos dias y buen Camino! Our connections to Spain continued. *El Camino de Santiago de Compostela*, or The Walk (or Way) of St. James, is a network of paths or routes across parts of Europe leading to Santiago de Compostela in northwestern Spain. In the Middle Ages, the tomb of St. James was the terminal point of pilgrimage for those who walked these routes. Today, thousands of walkers and bikers from all over the world and with many and varied motivations (sport, faith, nature, adventure, culture) make this trip. Some complete the entire length, and some opt for various routes that cover parts of the entire path. *El Camino* is a World Heritage Site.

During the months of April and May in 2018, my older brother, John, and some friends of his decided to travel the *Camino Primitivo* or *Original Camino*. They are experienced hikers and welcomed the challenge of traveling the oldest of the Camino de Santiago routes—from Oviedo, in Asturias, to Santiago de Compostela. This was the first Camino de Santiago trail when most of Spain was under Moorish control. The first stage of the Camino Primitivo, across the mountains, is one of the most rigorous of all Camino routes. When John traveled, as a favor to us and a gift to me requested of him by Jim, he carried prayer cards for James and left them at churches along the way, finally leaving one on the altar at the *Cathedral Santiago de Compostela*. Though Jim and I have not made the Camino, we vicariously walked it through John and his daily summaries from the emails and photos he sent us. Upon John's return we made a book for John and ourselves of all of his letters and photos. It was eighty-one pages, which makes my brother a prolific writer like his nephew! On the front cover of the book, we wrote: "*Senderista/Peregrino: John*" and also "*Caminando vicariamente: T & Jim*" ("Hiker/Pilgrim: John" and also "Walking vicariously: T & Jim").

***

Beth Adler Nesom has been one of my dearest friends since we were little girls. We officially met in second grade since we weren't in the same first grade classroom. Over sixty years' time she has helped me discern the true meaning of the word 'friendship'. Beth and Mike have three adult children, two daughters and a son. Our children grew up at the same time and are in the same age range. Andrew and our boys were pretty close in age. Between ten and fifteen years ago, their only son, Andrew, was deployed to Iraq - not once, but twice. Communication with him while he was gone and so far from home was extremely limited. Daily and/or as often as possible, Beth and Mike checked sites for updates while their son was in the thick of things, in a Bradley tank, and mostly not able to be in touch with home. After his tours, Andrew came home safely. He witnessed death and horror most of us will never see. Jim and I and our children followed news daily with Beth and Mike; I felt like we had a child in the Middle East … Meanwhile, James went to the Marshall Islands. He was assigned to a warless place of beauty far away from home. And he died … Flash forward to the present. We are sharing similar struggles and joys with outcomes that are now opposite between two of our children. Our shoes are on the other feet, so to speak. While I am happy for our family, I wish I could change a circumstance for these, our cherished friends. It is at this juncture that we recognize and have reinforced a valued lesson in the definition of friendship. We are people who journey through life together. *We share our life experiences—even though we cannot reduce the pain of the other or increase the joy of the other except by our presence.*

***

Shortly after James died, Julie Walsh from Hawaii called me. The voice over the phone said something like, "I'm with the University of Hawaii anthropology department. Originally I'm from Louisiana and went to LSU before your son, and I spoke with your son about him possibly attending school here in Hawaii for his masters when he finishes with World Teach. I used to be a volunteer in the Marshall Islands, too." Then she told me she'd heard of James' death and that she knew Miranda Powell and her family from being in the Marshalls, and she asked me if I would like to have copies of correspondence between James and her that she had filed in her office. Yes! Jim and I have always welcomed anything new—photos, papers, letters—anything we have never seen before relating to James. To this day, it amazes us when we see something new, something fresh. It is a bonus gift to see or learn something new about our son that we had not seen or known before. Julie was so kind. We exchanged contact information with each other, and we hung up. One contact we established was a friendship on Facebook to make it easy to exchange information. Then …

A completely separate and seemingly unrelated connection was made as Simone, our youngest, entered high school in Baton Rouge. One of her teachers, a particular favorite of hers and subsequently of ours, was a young man named Tim Hedrick. Tim taught religious studies and, within a few years of our getting to know him, decided to enter the priesthood. As he began studying, we gifted him with a Bible from our family and wished him well. We connected on Facebook so that we could follow his studies. Next …

Imagine my surprise when one day I saw communication between Tim Hedrick in NOLA and Julie Walsh from the RMI/Hawaii on Facebook! I contacted each to find out how they were connected, and I learned that Fr. Tim and Julie are cousins and that she is Tim's godmother. Again, worlds apart we have connections we would never have had if not for James. Julie came home to New Orleans for her cousin Tim's ordination. She drove to Baton Rouge for an entire afternoon just to meet us. We connected like old friends, talked for hours over cappuccino at our home, and are still connected with her and her family ... such a gift.

***

After James had settled into his hut on Arno Atoll, his *Baba* (who spoke little English) came to him one day with a bottle that had washed up a while back on the shore at Bikarej. He recognized that the bottle was something special since it had a note in it written in English, though he was not capable of reading it. There was a barely discernable return address. James helped his *Baba* write back to the man and woman who were from the United States and had written the note. They had traveled from New Mexico for a vacation to Mexico, which was the launch site of the bottle that had traveled for a year and nine months for over 6,000 miles by current all the way to the Marshalls! James and his island dad wrote back, telling them about the Marshalls and James' role there. James later received a return letter by mail telling his *Baba* and him that a package was on its way with some treats for the family in Bikarej. James was so, so excited about this connection. They waited and waited in their Marshallese world and in Marshallese time in which things happened whenever. The package could have been held up on the main island; who knew? Then James died. When we went to the memorials in the RMI, the package from the New Mexico couple had arrived and was presented to us. There were lures and other fishing gear for James and the residents of Arno. We left them all there for the Marshallese families on James' atoll.

We saw the names of these amazing, adventurous, generous people who lived in New Mexico from James' communication with us. But we had no phone number or discernable return address by the time the gifts were handed to us. When we got home to Louisiana, I was determined to find out who this couple was. I googled their names and found several in the area of Albuquerque. Then I did some digging, and I found out that one of them was a newspaper writer. I figured that maybe a person who would write a letter and put it in a bottle on a beach in Mexico could be a writer. That would make James and that person kindred spirits! So I called the newspaper in Albuquerque ... and the man who came to the phone was in fact the man who had written the note and put it in the trans-Pacific bottle. I told him about James and his death and was able to let him know that the gift was received and appreciated by the locals on Arno.

***

Lauren Pallotti Stumberg is an artist who was a volunteer when James was in the Marshalls. She stayed in touch with us after James died, and this connection inspired us to ask her to create a piece of art for our home. I had never commissioned anyone's services for art before, so this was exciting. She and I wrote back and forth through emails for months to get it just right. We wanted the emphasis of the art to be James on his island. We selected a photo as a basis for the piece. It was one sent to us by the RMI World Teach coordinators of James walking on the beach with two of his island brothers. Jim and I wanted additions around the perimeter of the work that were a snapshot of James' life at that time: life on the island; Philippians 4:7

(now a favorite Biblical quote of mine– "Peace is not a product of understanding why, but of trust—which surpasses all understanding"); James' writing; *EnaaJ Emman* ("It will be well"); fish; a traditional mat; an Arno map; the Bikarej lagoon; a stick chart; a basketball—encompassing many of the interests and facets of our son. The collage pictured here in these memoirs is the finished product that has hung over our fireplace since the day it arrived. Lauren named it *Lokanwa*, which literally means "the back side of the boat." She went on to explain to us that the word is used colloquially "to describe the sadness one feels as they watch loved ones sail away." We value it so.

***

Aylssa Cowell was Program Administrator of Waan Aelon in Majel (Canoes of the Marshall Islands) from 2010-2011. As of the writing of this book, she is a school counselor with eleven to nineteen-year-olds in Brunei Darussalam. As the beautiful art from Lauren was being prepared, Aylssa helped us understand stick charts and traditional navigation of the Marshallese waters, not only through the use of the stars, but also, uniquely, through the study of the swells and currents that are charted and utilized only in the RMI. Aylssa wrote, "When Europeans first arrived in the Marshall Islands, legend has it that they were so impressed by the ability of the Marshallese navigators that they tested their skills. They did this by putting the navigators within the hull of their ship so they were unable to see the stars, and they often changed course to try to confuse them. Needless to say, the Marshallese navigators were always able to find their way."

***

Before James left, I made him a book with envelopes per page per month for his year in the Marshall Islands. Since we knew that personal contact with James would be challenging and sporadic at best, I wanted him to know that we were thinking of him all the time. In each envelope there were notes, greeting cards for the season or holiday (Christmas, New Year's, Easter), prayer cards, and sometimes money (such as during the month of December for his birthday). When we arrived in his little room in Bikarej at the time of the memorials, I found the book right away. James was a rule-follower at his very core, and he was self-disciplined. It was November when he disappeared. Now it was December when we stood in his little space of a room. The money for his December month was still there in the as-yet-unopened envelope.

***

Right before James left home for the Marshalls, he and I talked about his faith. James was in a questioning and exploring time in his life. With intentionality, he studied various faiths and ruled some in and/or out based on his study. He had attended Mass during most of the years of his life, and then he stopped. He was an altar server past the time many of his friends stopped serving. He wondered why and how with regard to many facets of life. He prayed and then he didn't and then he did … James was more and more open with me and calm in discussions as he matured. In his letters, he mentioned to me when he went to church in the Marshalls, because he wanted to make me happy (note: letter dated 08/27/10). During one of our last lengthy conversations together before he left home for the Marshalls (and I still wonder what possessed me to ask this), I asked James this question, "If something happens to you (not thinking it really would) and you meet God and He asks you if you believe in Him, what would you say?" He responded, "Of course!" He went on to say he just wasn't sure yet what he understood; he wanted to know more. But he said he would respond yes to God if he met Him. That brought me some comfort.

As providential as that conversation was, a final communication between Michelle and James seems equally or more comforting and providential to me. James' siblings took him out for a final night all together in Baton Rouge just before he left. While they were out, his oldest sister gave him a card that read:

As I think of all of the things that I want for you while you are on this trip, some very important things come to mind first:

- *I hope you stay healthy.*
- *I hope you are surrounded by people who are, at their core, kind and loving and sympathetic.*
- *I hope you never feel scared or hopeless.*
- *I hope that, if you do feel scared or hopeless about anything while you are away, you are surrounded by people who will help you to feel the way you know your family here would.*
- *I hope you call upon your faith to help you through all times, be them trying, amazing, upsetting, or joyful.*

I love you and will be thinking of you <u>every day</u>.

We found this note in the book of notes and cards I made for James to open monthly during the year he was to be in the RMI. Michelle's note was tucked into the November slot with what he had opened from me. Michelle had given that card to him before he left, but it seems James had read that beautiful note as recently as the month he died. I'd like to think he moved it forward each month in the book because it meant that much to him. What joy this thought gives me as his mother.

Just before sending this memoir off to begin the publishing process, I wrote a note to Michelle in March 2020 to tell her I'd found her note tucked into the November page of the book I'd given James. Michelle replied, "I know! I saw that he saved the note I wrote him when I found it in the November section of the book you made him while we were in the Marshalls gathering his things. When we were in the Marshalls and I reread my words after everything had happened, I found them very sobering and prophetic! I also realized, while I read them, that I meant those words so completely and genuinely. And, I believe that he was probably scared at some point during the climax of that event, even if for a brief moment, and I believe with my whole heart that he was in fact surrounded by people who would have protected him and who tried to protect him to the best of their abilities, just as his family would have. I hope in some small way, even if it was subconscious, that he felt us all in his final moments."

***

James was thinking about his future and wrote to us that he was uncertain what the next chapter of his journey would be (note: letter dated 11/02/10). Graduate school possibly in Hawaii? Another year with World Teach? Another volunteer year with a different organization? Teaching in Spain? Travel? These are eight little words written to me by James on 11/2/2010—just 23 days before he and the others on his boat were lost, *"Ma—don't plan just yet for my return."*

…

25

Of Joy, Thanksgiving, & Peace

So how did we reach this place of peace? Is it possible to swing full circle through:

JOY on November 24, 2010 when all seemed well with the world to
SHOCK on November 25, 2010 to
DESPAIR and DEEP SADNESS in the following months to
GRIEF with erratic, gradually increasing
RELIEF from time to time over ten years to
JOY *again*—and increasing
HOPE now—here in 2020?

Ten grief-filled and sometimes brutal years of striving for faithful acceptance, gradual deepening of faith and hope because of James (that may not have otherwise occurred to such a degree), noticing the beauty and true gifts of family and friends who travel this journey so selflessly, acknowledging God-winks that happen around us if and when we just have eyes and ears to notice, admitting more and more willingly our gratitude, experiencing the "co-incidences" that clearly are not merely just coincidences, and so much more … all of this coalesces to create our joy even as we wait for it all to be complete. Our grief is not entirely gone; I believe it never will be entirely gone.

Throughout our journey, we are more and more consoled by our belief that we will eventually be together with God and again with James and others who left this Earth before us … as well as those who have not yet arrived here. To me, this presence of all of these members is the definition of the Communion of Saints in eternity now.

To everyone who contributed to *Jambos With James*, I am more grateful than I can express for your openness and generosity in sharing yourselves and your time to connect with our family. To anyone who reads this memoir, thank you for sharing this remembrance and tribute to our son, James. Sharing his life has helped me and has taught me much. This process is life-long, but it's worth sharing at this juncture with those going through similar losses.

Maybe you have lost a child. Maybe you have lost a dear friend. Maybe you have lost a sibling. In my grief group, we have each lost a child—whether to murder, drowning, drugs, suicide, wrecks, cancer … the list goes on. There is not a right or wrong way to do this … this grief. But I believe we who grieve have more in common than not. A community of support is a gift. That may not fall right into place, and it may need to be kindled. But the effort is worth it. We are all on this journey together; we share our travels.

***

Here, providentially, is the end of the last letter James ever penned to us - just days before he died.

His very last line ever written to us proclaimed,

"You may know this song as JOY to the World!"

(note: letter dated 11/03/10)

We are grateful.

PART TWO

Abigail Mercer
Abigail Sherlin
Adam Delaune & family
Adam Ducoing
Adam Solino
Adelaide Jewel
Adrien & Trouty Larroque
Adrienne (Persac) Taylor w/ Bryan & David
Aislynn Herrera
Alcia (Lemoine) & Randy Plaisance
Alex ... in Spain
Alex Grush
Alexandra Goodwin
Ali & Brian Hymel & family
Ali (Becnel) & Adam Solino
Aline Blanchard
Alli Matthis
Allison Brown
Allyce Baudier Lemon & Xander James & Asher
Allyson Thibodeaux
Allysse Dambach
Alycia Domma & Bryan Ford
Alyssa Cowell
Amanda (Cornelius) & Tim Parker & family
Amanda Colon
André Cornelius family
André Nicole deBrueys Cardinale
Andrea & Luke Bacino
Andrea Wesner
Andree (Trahant), Ray, & Ava Price
Andrew & Courtney Nesom
Andrew Cambus
Andrew Herpich
Andrew Rayner
Andy Burka
Andy DuGan
Andy Florek
Angela & Jon Bajon
Angela & Randy Green
Angela Bajon's same-day surgery staff (BR General)
Angela Falgoust
Angela Saunders
Angela Weisse
Angelina Jimenez
Ann & Tim Riley w/ Ann, Sean, & Evan
Ann & Tommy Bernard
Anna & Mike Murphy
Ana (Fernandez) Payne & family
Anna (Little) & Jed Robinson & Madeline
Anna ... in Spain
Anna Arno
Anna Linares McCann & family
Anna Nesom
Anna Wall
Anna Zelinsky

Anne & Fred Trap
Anne & Mark Wojna w/ Emily & Elizabeth
Anne Avant
Anne K. Bailey Liebowitz
Annelie Smolenski & family
Annie Himmelsteib
Antonio Carbajal
Anwel Ninne (lost on boat)
Arnold Alper family
Aryn Kamener
Ashleigh Tassin
Aylssa Cowell
Baby Kiako & her unborn child (lost on boat)
Barbara & Dan Wanko
Barbara & Maurice O'Rourk
Barbara & Umpy Brown
Barbara Holden
Barbara Kearny
Batista family
Bea & Bill Brockman
Becky Allen
Beckye Taylor
Belinda Fenner
Beth (Adler) & Mike Nesom
Betsy & Jay Dahlke w/ Greg & Scott
Betty *Dardenne) & Steve Backstrom
Betty Claire & Mel Rabalais & family
Beverly Schroeder
Bianca (Spain)
Bill Conway
Billie Andersson
Bishop Greg Aymond - NOLA
Blake Patrick McCrary
Bo & Vicki Richardson w/ Bo
Bobby & Courtney (Rooney) Helou
Bonnie & Joe Torres w/ Jessica & Ryan
Boo Flynn
Brady Lemon & Xander James & Asher
Brandi & Michael Matthis, Alli & Heather
Brandon & Rebekah Allen
Brenda & Andy Redpath & family
Brenda & Deacon Drea Capaci
Brenda & Richard Follette & family
Brett & Charles Ward
Brett Bordelon
Brian, Emily, & Caroline McCrary
Brianna LeBlanc
Brittany (preschool)
Brittany Collins
Brittany Jewell
Brittany Sherlin & family
Brooke Payne
Brother Chris Sweeney

Brother Louis Couvillon
Brother Martin @ New Orleans - Class of 2006
Brother Martin High School @ New Orleans
Brother Ray Hebert
Bruce & Paulette Curson w/ Cory
Bruce & Shelley Alexander
Bryce Luquet
Bungalow Bill
Callie Romero
Camy (Spain)
Candice Maddison Guavis
Carmen (Linares) Simons
Carol & Roger Peck
Carol (O'Keefe) Barback
Carol Maracolis
Carol Poche'
Carol Regan & family
Caroline Vanek
Carolyn Haar & family
Carolyn Murphy
Carrie Reed
Cassy Davis
Cathy & Bill Marchese
Cathy (Levert) McKinnon
Cathy Selleck Carmouche
Caz Hologa
Charlene Stern & family
Charles Yost
Charley & Charles Strickland
Charlie Sartain
Charlotte & John Little w/Anna, Emily, Laura, & Katy
Charlotte (Kidd) Ray
Charlotte Bracey
Charlotte Hymel & Mike Bracey
Chaz (preschool)
Cheryl Frost
Chisty & Fed Wild
Chris & Holly Rigol & family
Chris (Layman) & Herb O'Niell
Chris Daigle
Chris Haar
Chris Vizzini
Chrissy White
Christina Stenhouse
Christine & Bryan Bordelon
Christine Faureau
Christopher Vicknair
Christy & Billy Kirkikis & family
Chuck Branton
Chuck Luquet & family
Chuck Wimberly
Cindy Stumpf
Cissy Yakelis
Claire & Steve Wilson
Clara Richmond & family
Clare Jones
Claude & Suzi Fristoe
Clementine White
Coach Curson & family
Cody Luquet

Colette & Ramsey Reimers & Hotel Robert Reimers staff (RMI)
Connie Maranto
Connie Superneau
Corrie Borné
Cory Lemon
Cottage Care of Baton Rouge
Courtney (Green) & Jesse Hayes
Craig Clement
Cyndie (Eicher) & Steve Bergeron
Cynthia & Brad Wallace
Daniel Blanchard & family
Daniel Erazo
Daniel Moran
Darlene & Mike Finnan
Darlene Woodford & Come, Lord Jesus prayer group @ St. Alosyius Church
Darsi (Taylor) Mellor
David Chicoine
David Nichomoff
David Pierson
Dawn, Eddie, & Adam Tessmer
Deacon Drea & Brenda Capaci
Deacon Jack Young
Debbie & Bob Delaune
Debbie & Mark Wamsley
Debbie & Pete Alongia
Debbie Perrone & family
Debbie, Bob, Adam, & Kailen Delaune
Debora (Fuchs) Porazzi
Deborah Hallen & Paul Zelinsky
Debra Sarver w/ Tyler
Denise Boyce
Denise deBrueys
Denise Nagim
Denni Mulroy
Desiree Costillo
Diane & Bl& O'Connor
Diane & Ed Neumannw/ Michael
Diane & Rich Gonzalez
Diane Neumann
Dianne (Dragon) Herbert
Dolly Duplantier w/ Matt & family
Donald Beale
Donna & David Fargason & family
Donna & Jeff Hologa
Donna & Jim Bordelon w/ Jason
Donna Olinde
Doris Brignac
Dottie & Jim Richard
Doug & Rachel Alongia & Finn,
Dougie James, Art, & Maggie
Dougie James Alongia
Duane Superneau
Dutch Prager
Dwayne Broussard & family
Edie Boudreaux
Elaine Cucinello & family

Eleanor Morris
Elena Ronquillo
Elise Orellana
Elissa McKenzie & family
Elissa Minet Fuchs
Elizabeth & Alan Schroeder & family
Elizabeth (preschool)
Eloise & Jim Foreman
Emily Balls
Emily Eck
Emily Mayo
Emily Perkins
Emma Claire Bajon
Emmett Waggenspack
Emmy & Craig Clement
Eric McCrary & Elizabeth
Erica Moore
Erik Watnik
Erin & Chuck Coolidge
Erin (Nesom) Jackson
Erin Candilora
Erin Elizabeth Mikulak
Erin McLean
Erin Uzee
Erin Wimberly
Esther & John Gin
Evelyn "Mom" (Green) Utke
Evey & Augie Berner
Faith Camet-Andries & family
Faith Wakefield
Fr. Cassian (Todd) Derbes
Fr. Donald Blanchard
Fr. Eddie Martin
Fr. Eustace Hermes
Fr. Gerald Burns
Fr. Howard Hall
Fr. Jose Lavastida
Fr. Josh Johnson
Fr. Randy Cuevas
Fr. Ross Romero
Fr. Than Vu
Fr. Tim Hedrick
Fran & Jack Hannan & family
Fran Marmillion
Frances & Gary McConnell
Frances & Johnny Linker & family
Frances Fakouri
Frank France
Friends of James from Baton Rouge
Friends of James from Brother Martin High School
Friends of James from Kenner & NOLA
Friends of James from LSU
Friends of James from our Kenner neighborhood
Friends of James from Seven Oaks preschool years
Friends of James from St. Elizabeth Ann Seton elementary years
Friends of James from the RMI
Gail & Darryl Bordelon
Gary & Kay Giepert & family
Gay Hebert
Genie Kleinpeter Silva
George & Martha Jane Grammer
George & Nell Leggio
Gerald Tullier
Ginger & Jeff Cardinale
Ginger Cardinale's Church of Christ friends from Morgan City
Glenda Pollard
Gloria & Lewis Chairs
Grace Doll
Grace Perez & family
Greg & Charlene Raymond
Greg Hurst
Greg Rando
Greyson Bordelon
Guadalupe & Luis Erazo
Gwen & Dan Schaneville
Hannah Page
Hannah Thornton-Smith
Hanneke Van Dyke
Helen Claire Sievers
Helen Claire, Shawn, & World Teach Staff
Hilda & Walter Krousel
Hope & Nathan Harrison & family
Hope Himel-Benson & family
Hope, Gary, & Grant Benson
Ian Brown
Ida & Dan Moran
Ingrid Ahlgren
Island family - Salem (Jeral)
Island family - Frank (Bulijul)
Island family - Taji
Island family - Katije
Island family - Taki
Island family - David (Biku)
Island family - Wiskey
Island family - Clora
Island family - Romita
Island family - Billa
Island family - Rod
Island family - Rickji
Island summer camper - Yoho
Island summer camper - Elwina
Island summer camper - Jojo
Island summer camper - Jejee
Island summer camper - J.R.
Island summer camper - Wiff
Island summer camper - Almond Joy
Ivy & Bruce Phillips
Jack Morin
Jackie & Dick Upton
Jackie Zeller
Jackie-Sue & Sam Scelfo
James & Tammie Domma
James Catalano
James Clifton deBrueys
James Henderson & family
Jan & Charlie Spansel & family
Jan Dimattia
Jan McCurdy
Jan Radford
Jane & Bill Goldring
Jane & Tom Hughes
Jane Barney
Jane St. Pierre
Jane Villarrubia & family
Janet Bohannan & family
Janet Landry
Janet Pananos
Janice & George Papale
Janie & John Loftus w/ Kendall
Janie (Munson) Berg & family
Janis & Patrick Eck
Janis Landry
Jason & Karla Rigol & family
Jason (preschool)
Jason Papale
Jean & Ed Smith
Jeanne & David James & family
Jeanne (Fourrier) Eggart
Jeanne Anne (Grass) & Phil LaRose
Jeannette & Celestin Kasongo & family
Jeannette & Mike Rolfsen
Jeannie & Gary Vega & family
Jeannie Rappold w/ Matt & John
Jeannie Ruda & family
Jeannine (Gerald) Schutte
Jen (Caluda) Davis
Jen Bristow
Jennifer & Michael Mayer
Jennifer Golden
Jenny & Walter Morales & family
Jerry & Debbie Taylor & family
Jess Smith
Jessica (Bacino) Cerruti & family
Jessica Comeaux Ehlers
Jessica LeJeune
Jessica Thompson
Jill & Chris Abadie
Jillian T. Crawford
Jim "Coach" Gilbert
Jim & Gwen Henderson & family
Jim & Mary T (Heffron) deBrueys
JJ Alcantara
Joan & Bill Bailey
Joan & Charlie Freel
Joan Babin
Joan Bailey
Joan Danos
Joan Kathman
Joann (LaPorte) Lappin
Joanna Lemoine
Joanne & Cooper Roberts
Joanne Baston & family
Joe & Margaret Tusa
JoEllyn (Abadie) & Billy Gallman w/ Heather, Lindsay, & family
Joey Chauffe
Johji & Keiko Kurodo
John & Kathy Crabtree
John & Susan Heffron
John Gambino
John Sebastian & André Cardinale & Nizza
John Sutherland & Susan (Gallagher) Heffrom
John Vanek & family
John Vizzini
Jolie (Ranzino) & Carey Messina
Jolie, Brent, & Brianna LeBlanc & family
Jon & Angela Bajon
Jon Parker & Simone Bajon & Luke & Emma
Jonathan Caluda
Joni & Tommy Diaz w/ Grant McClure
Jordan Rodgers
Jordon Dupont
Jorje & Ruby Rodriguez & family
Jose' Perez
Joseph Delaune
Joseph Hansen
Joy & Vic Weston
Joyce & Bill Kilshaw
Joyce & Dyer Lafleur
Joyce & Steve DuGan
Joyce Moolekamp
Juban's Restaurant
Judy & Fred Derbes & family
Judy (from NO Cooking Experience)
Judy (Lindsey) & Mike Determann
Judy Bonnecaze
Judy Danos
Judy Hogg Mills
Julie Ann Walsh & Miles & Max
Justine & Stephen Gilbert
Kalan Warrick
Karen & Danny Sansovich & Lauren/Christian & Emily
Karen & David Scariano
Karen & Jim Reichard
Karen & Tom Hickey
Karen Edwards
Karen Nugent
Karin & Tom Gerace
Kathryn Lewis & family
Kathy DelRio & family
Kathy & David Waltemath & family
Kathy & Jack Williams w/Lauren & Kathryn
Kathy & Jerry Meunier
Kathy & Sal Faia & family
Kathy & Wayne Vizzini
Kathy Meares
Katie (Caluda) Santos
Katie (Finberg) & Joe Fotofili & families

Katie Saer
Katy Dubus
Kay & Howard McKissack
Kay (Haase) Aucoin
Kay McKissack
Kaylee Crousillac
Keeli Cahalan
Kehoe-France Faculty @ Metairie, LA
Kempton Baldridge
Ken & Cassie Price
Ken Hagberg & family
Ken Thevenet
Kenny Spellman
Kevin & Chris Cahalan & family
Kevin & Kristen Moran
Kevin Hulin
Kevin Martin Heffron
Kevin Vega
Kim & Bobby Shackleton & family
Kim & David Gambel & family
Kim & Ricky Spindel & family
Kim Bowman
Kim Vinci
Kimberly Blanton Hafner
Kiotak Abitlom Joream (lost on boat)
Kit Reilly
Kitty Cleveland & family
Kotler family
Kristen & Darren Horn
Kristie Larson
Kristina Bramwell
Kristy Dupree
Lara Farina
Laura Palmintier
Laura Sundblad
Lauren (Beaulieu) Williams
Lauren (Landry) & Michael Cashio
Lauren Pallotta Stumberg
Lauren Schayot
Laurie & Gerry Catalano
Laurie Prange
Lee & Judy Miller
Leila & David Dragon
Lelia & Joe Prange
Lena Haynes
Lenny & Gayle Betzer
Lenore Wamsley
Leon Knight
Lila Seymour
Linda & Bob McCrary & family
Linda & Johny Vanek
Linda & Rob Christensen
Linda Grush
Linda Harvison
Linda Lane
Linda Malone
Lindsey Ryan
Lindsey Schayot
Line Ma
Lisa & Johnny Veron
Lisa Disney
Lisa Mongrue
Lisette & Dan Borné
Liz Behsudi
Liz Duplantier
Liz Seiter
Liz Zelaya
Lloyd & Diana LeBlanc
Lois Hernandez
Louise & Allen Kincannon & Louise's 4th grade class
Louise Prosser
Lourdes (Delgado) & Jim Stoddard & family
Luis Erazo
Luke Philip Bajon
Luke Strachan
Lury & Antoine Ignizio & family
Lyn Ace
Lynn & Bill Hartley
Lynn & Henry Carville
Lynn & Steve Meade
Lynn (Bani) Carville
Lynn (Hagen) & Denny Otillio
Lynn (LaPorte) & Stew Samuels
Madeline & Bill von Almen & family
Mae & Thomas LaPorte & family
Maibel Hurtado & family
Mandy Doyle Cohen
Maranda Blount
Marcy (Boucher) & Bill Stairwalt & family
Marcy Boucher
Margaret & Manny Pineda
Margaret Wilkerson
Margie O'Connor
Maria Guarnieri Vranic
Maria Henson
Maria Woodyear Bowen & family
Marian & Charles Gant
Marian Guarnieri
Marianne Cardwell
Maribel Hurtado & family
Marilyn Meyn
Marilyn Spadora & family
Marisol Valladares & Elena
Mark & Adele Morin, with Jack, Nick, & Max
Mark Kirkikis
Marley Bordelon
Martha (Beck) & Mark Upton
Martha Carbajal
Mary & Craig Roussell
Mary & Howard Penton
Mary & Phil Pitts
Mary & Rodney Guillot
Mary & Sean Dowd w/ Jennifer, Maddie, & Ryan
Mary & Spencer Schayot w/ Lauren & Lindsey
Mary & Watson Tebo & family
Mary (Reeks) & Phil Arno & family
Mary Alice (Beck) & Richard Rathe w/ Christine, Katie, John & Ann
Mary Ann Hotstream
Mary Barbara & C.W. Kinchen
Mary Beth & John Hagberg & family
Mary Beth Kling
Mary Carol McNamara
Mary Dawson
Mary Eleanor (Lepine) & Lanny Harris
Mary Ellen & Jere Price
Mary Helen & Mike Miller w/ Gordon & &rew
Mary Karam & family
Mary Landrieu
Mary Liz (Henderson) & David Paley w/ Alex & Lauren
Mary Lou (Verges) & Don Barron w/ Kelli
Mary Lu & Glenn Penton & family
Mary Lynn & Mike Garin
Mary Lynn Langlois
Mary Mang & family
Mary Martha Carbajal
Mary Peters
Mary Reagan (Moran) & Jordan Jacob
Mary T (Heffron) & Jim deBrueys
Mary T (White) & Norman Clifton Heffron
Mary T Madison McCrary
Mary T Michelle deBrueys McCrary
Masika Kaniki & family
Masika Kaniki w/Bienveni, Clovis, Claude, Clement, Claudine, David, & Esther
Matt & Natasha Crousillac
Matt Fallon
Matthew Bordelon
Max Hoegh
Max Materne
Max Morin
Maxine Larroquette
Mayor Mike Yenni
Meaghan Tower
Meghan Dombourian
Melanie & Craig Vitrano
Melanie & Gary Vega
Meli & Ralph Melancon
Melissa (Gibb) & Ward Muir
Melissa Giles Bell & family
Melody Kitchen & family
Melvina White
Mercedes & Jim Dore'
Merideth Piggott-Tooke
Mia & Jeff James
Michael & Jennifer Mayer
Michael & Sara Spadora & family
Michael Buccola
Michael Franco & family
Michael Kuhn & Maria Eliott
Michael Powell
Michael Wattingney
Michaelyn Bellelo
Michell & Scott Rabalais
Michelle & Patrick McCrary
Michelle & Phil Waguespack
Michelle (Russo) & Chris Fischer & family
Mickey Adler & family
Mike & Beth Nesom
Mike & Rosa Dunn
Mike Reif & family
Millie & Billy Blanchard
Mindy & Kevin Colomb
Miranda & Ryan Powell & family
Missy Jones
Mollee (Di Benedetto) & Gary Vicknair
Mollee, Kurt, & Gary Vicknair
Mona & David Strauss
Monica & Frank Fazio & family
Monica Inthavong
Monica Montgomery
Monique & Lou Wesner & family
Moo Turner Svenson
Moselina Reyes
Nancy & David Fourrier
Nancy (Klug) & John Loughner
Nancy (LeBlanc) & Beau Bondy & family
Nancy Mumford
Nancy Savoy
Nat & Deacon Dan Cordes
Natalie Nimmer
Nathan Roy
Nell McAnelly
Nick Morin
Nicky & Garland Bourgeois
Nicole Mercado
Nicole Vizzini
Nika Wase
Nikki (Fuchs) Vidrine
Nizza Katerina Cardinale
Noa Silva
NOLA Metropolitan Crime Commission
Our Lady of Mercy Catholic Church @ Baton Rouge
Paige Harris
Paige Morrison & family
Pam & John Richard w/ Scott & PatrickEric & Lisa McCrary
Pat & Jim Pettus
Pat & John Crabtree
Pat & Ken Nieto
Pat Betz w/Ginger Robertson
Pat Veazey
Patrice (Comeaux) & Richard Ellis
Patricia Briggs Gordon

Patricia Henderson & Braonton family
Patricia Thomas
Patrick John & Michelle McCrary & Maddie, Blake, & Samantha
Patsy & Ken Domma
Patsy & Ron Richmond
Patte Keogh
Pattu & Darryl Bonura
Patty Glazer
Paul & Beth Geohegan
Paul Siragusa & family
Paul Spadora & family
Paula (Heffron) & Danny Moran
Paula Moll & family
Pearl & Bob Squires
Peggy Rivault
Pele Levy-Strauss
Peter Alongia
Peter Barnes
Peter Haar
Peter Rudiak-Gould
Phil D'Antonio
Phillip McManaman family
Phyllis Parks
Rabenhorst Funeral Home
Rachel & Doug Alongia & Finn, Dougie James, Art, & Maggie
Rachel (Grammer) Peterson
Rachel Easley
Rachelle & Josh Berkley
Rachelle & Norman Albright
Rafe & Val Goyeneche
Ramona & David Strauss
Randi & Jordan Bergeron & family
Ray Dawson
Ray Yakelis
Reagan Houston & family
Rebecca Nichomoff
Remell & Craig Goodwin w/ Alexandra, Remy, & Ridge
Rene & Nancy Nguyen w/ Vickie
Renee Landrieu & Stephanie Harkness
Ret Ron
Rhett & Jessica Ward
Rhodes & Kelsey Moran
Rhonda Melancon & family
Richard & Cathy Carmouche
Richard Carmen
Richard Tulley
Rine Ma
Rob (in Spain)
Robert Frank
Rose & Mark Herzog
Rose Marie & Johnny Fife
Rose Mary (Romero) & Bill Mize
Rossana & David Hnatyshyn
Roz Tabor
Ruth & John Pace
Ryan Coulon
Ryan Dendinger
Ryan Dennington

Ryan Derbes & family
Ryan Gautreaux
Ryan Torres
Sally & Mike Czerwinski & family
Sally (Pettit) Wimberly
Samantha Chapman
Samantha James McCrary
Sandra Cyprian w/ Jasmine Mackson & family
Sandy & Greg Brown
Sara Sisto
Sarah & Stephen Leard
Sarah (Heffron), Justin, Rebecca, & David Nichomoff
Scott &rew Melancon
Scott Hogan
Scott Melancon
Sean Cassidy
Sean Lou
Senator David Vitter
Serena & Kirk Jones
Seven Oaks Academy @ Kenner
Shalisa Bynam
Shane Schexnaydre
Shannon McKernan & family
Shari & Henry Pere'
Sheila & Bob Butler
Shelley & David Chryssoverges
Sheri Gillio
Sheridan & Aimee Moran
Shirley & Ed Neumann
Shivaun Tessier Davis
Sidney Blakemore
Sima Zadeh
Simon & Shelly Cornelius & family
Simone & Jon Bajon
Simone Claire deBrueys Bajon
Sissy & Ralph Stephens
Sonya Miller
Sr. Bert & the Sisters of St. Joseph
Sr. Camille Anne Campbell
Sr. Carolyn Brady
Sr. Cynthia Sabathier
Sr. Diane Hooley
Sr. Dianne Fanguy
Sr. Janet Franklin
Sr. Joan Laplace
Sr. Shirley Vaughn
St. Aloysius Bereavement Committee
St. Aloysius Catholic Church @ Baton Rouge, Louisiana
St. Aloysius Catholic School @ Baton Rouge, Louisiana
St. Elizabeth Ann Seton Catholic School @ Kenner, Louisisna
St. Jerome Catholic Church @ Kenner, Louisiana
St. Joseph's Academy @ Baton Rouge - Class of 1972
St. Paul's United Methodist Church @ Kerrville, TX

Stacey Keaton
Stan Brien & Laura Brown w/ Jack Brien
Stepahnie (Rogers) & Ed Fike w/Haley
Stephanie Bueche
Stephanie Coxe
Stephanie McCutcheon
Stephen Leard
Steven Glenn deBrueys
Stuart Lovinggood
Sue & Tom Deavers
Sue Day & family
Susan & Joe Locascio & family
Susan & Rob L'Hoste
Susan (Gilbert) & Joe Locascio & family
Susan (Lafleur) & Rick Parker
Susan Blanchard
Susan Bosche' & family
Susan deJong
Susan Delaune
Susan Hancock & family
Susan Heffron
Susan Piggott & Michael Tooke
Susan Wegmann
Susan Woodward
Suse & Tom Deavers
Susie & Mike Cresap
Suzanne & Michael Pettus & family
Suzanne Kirkwood & family
Suzette Herpich & family
Suzy & Andy Adler & Allie, Katie, & Matt
Sybil Roussell
Sydnie Shackleton
Sylvia Finger
Tamara (Greenstone) & John Alefaio
Tara (Carlini) LeBlanc & family
Teresa & Pat Leard
Terry & Joe Abramowicz
Theresa & Mitch Leger
Therese & Bob Barry
Therese, Wesley, & Julie McCann
Tiffany & John Meek & family
Tina & Michael Cashio
Tina (Bergeron) Hector
Tish & Ronnie Gerald
Tito & Susana Rey & family
Todd Mulroy
Tommy Gerace
Tommy Mitchell
Tony Ma
Tracey Barhorst
Traci (Roussell) & Marc Gerald & family
Trina Derenbecker
Trio #1 - James, Steven, & Simone
Trio #2 - James, Chris, & Brett
Trio #3 - James, Doug, & Luis

Trish (Luquet) & Paul Otts & family
Trisha Brown
Tyler Luquet
Tyler Rheams
Valerie (Lazickas) & Provan Crump
Vance & Deborah Gibbs
Vera & George Marse
Veronica Wase
Vicki (Montalbano) & David Brewer
Victoria Heffron & Stepehn Price
Vincent Scelfo
Vivian Guarisco
Walter LeBon
Warren & Mary Henderson deBrueys
Wendy Gelpi & family
Will Kirkikis
Xander James Lemon
Yi Dai
Yvette (Cornelius) & Jan Jircik & family

PART THREE

Part Three of this book contains James' letters.
They were all handwritten by him, and I have typed them out just as he wrote and spelled them.

8/6/10

Dear Mom, Pop, and all you chirren,

"Yokwe Kom!" (Or "hello everybody") (and I misspelled)

It's 8/5/10 at home.

I'm currently in a classroom doing practice teaching with some students at Delap Elementary School. I am working with 4 other soon-to-be-teachers. We all lesson planned together for second/third graders at "World Teach Summer Camp." I talked about clues and emotions, creating sentences and singing songs along the way. I received some of y'all letters and e-mails, so I appreciate that. I am going to try to get on the Internet soon.

I met my principal the other day when I received my Marshall Islands Teacher Certification. His name is Alex, and he appears very excited. I will be teaching only English, though he already told me I could do whatever I wanted, even asking me to set up workshops. He also told me "I am not your boss, and you are not mine. We are a team!" I think he and I will be pretty good friends. I am the first ever "ribelle" (essentially "gringo"; literally "one who wears clothes") to volunteer on Bikarej. I think I might be more isolated than previously thought. Woo!

Majuro is a pretty cool place, I suppose. It is cool that the people here are so kind and trusting. For example, I have hitchhiked down the one road pretty often and at pretty good distances. Majuro downtown is about 13 miles from Ajeltake, which is where I am living with my 32 other roommates.

Ajeltake Elementary is where we do communal cooking, cleaning, and lesson planning. Everyone has been great so far, even though we all share two bathrooms (with bucket flush!) and three showers (with buckets!). The stars are great at night ... It's not raining. It rained in bursts all day and night. And when it's not raining, it is really hot and humid. I'm slowly getting used to it. Taking a dip in the lagoon feels great. I've seen some beautiful coral and a sunken plane from the 50s. And the fish are just crazy. When I get to Arno Atoll, I should even better snorkeling. (However, my dive shoes were "borrowed" ... . Everything around here is shared and together, so someone decided to take my shoes. If possible, could you send me a size 13 pair? They have them at the dive shop on Government! I love you! Also, I can use the clippers out the movies; plus, they are the best protection against the coral.

I never finished talking about Majuro at all, so I will now. Stuff is expensive here, and there is a lot of pollution. All I can say that I'm glad I got my tetanus booster - a lot of rusted metal. I am also getting used to eating pretty much the same things every day (but fresh sashimi and breadfruit with coconut sauce is delicious everyday). There are about four or five bars, a few groceries, and a couple hotels. It is very slow and calm here, sometimes almost too much. I don't know what some of these people do all day - I've seen many just sit in the shade or play basketball once it is dark. And they like to get drunk at night. And not in a good way. But, again, everyone is very nice, and the kids here are awesome.

By the way, my summer camp kids are named Yoho, Elwina (the girls), Jojo, Jejee, J.R., Wiff, and Almond Joy.

I have been studying some Marshallese, prepping for my boat trip (and trying to get to my island proper), and just hanging out with these 32 new friends that I may never see again. Tonight, we celebrate the end of the practicum with a BBQ and karaoke, so it should be a hoot. I believe I leave August 15th-18th ... but who knows. I am going to try to contact before I set off, but I will give this "list" of things I may like (:)): snorkel booties size 13 (goes up to the ankle), a headlamp, Dr. Suess books/other children's books, AA batteries, children's puzzles/posters, and whatever else you deem fit. Thanks for anything and everything! [In the left margin of James' letter, he wrote three notes: *Oh, and TONYS! *Oh, and PENS! *Oh, and DRIED FRUIT!]

Steven should be leaving soon, how is that going? And André and John – what's the word? Is Simone at André's? How was Madison's birthday (tell her I said "Haaapppyyy BirthDAAAY!), and all the McCrary's? How's my Grandmother? Is she behaving? And Grandpa and Aunt D? And, of course, the parentals. How y'all? I will try to keep yall updated hen I can.

Oh, and I am planning for the children of Bikarej to say yall – it will be the most Southern state in the Pacific.

I love all yall, and I hope everything is going well…And that you can read my handwriting. I will update you when I can.

Yokwe, & Love,

James

Bar lo kom! [See you later!]

***

8/6/10

Dear Jess, Luis, and any and all who read this letter!

It's 8/5 for you.

Yokwe! How y'all all doin? I am currently in Majuro, the capital of the RMI, in Delap Elementary School. I just finished teaching at "World Teach Summer Camp," where in order to get butts in seats, we enticed children with candy. In the Marshalls, that's apparently not a creepy thing. It's pretty cool here-great snorkeling and nice people. Everyone hitchhikes everywhere (on the one road …), going miles out of the way to bring you home. There is a lot of pollution and diabetes-a lot of adults have missing legs. Teaching, thus far, has been fun. I go to Bikarej on August 15-19 … or maybe later. It's very unsure. Fun fact: boats don't usually go where I am going, so they are trying to figure out if they can charter a boat for me. Bikarej has a population of 200, with 60 kids at the school (which is only three rooms). I am teaching primarily English class, and I have 3 lagoons. And that is all I still know. Oh, and I am the only "ribelle" (a.k.a. "gringo") that will be there (and that has ever lived there).

All the World Teach people are cool, and all of us live in a non-a/c school building in a town called Ajeltake. I am learning the art of bucket showers. I used about 2 gallons. I am trying to get it lower. Dude, the snorkeling rocks, and I bet it'll be better once I'm in the Arno Atoll. I've seen a lot of awesome coral and a sunken plane. And all sorts of fish. It's so beautiful under this water. It's so blue and crisp.

Just, I got your postcard. Thanks for the preemptive strike! Show the letter to all the chirren back at home.

When I'm on my island, I may be more removed than I originally thought, as no boats go to Bikarej often. So letters will be slow, and packages slower.

I hope all y'all are good. Have a beer at my favorite haunt for me ... maybe two. I send all my love to you crazy kids. Let me know how you are doing. I'm going to send letters at random, so be aware.

And sorry the letter is not longer; just finished writing one to my parents. I'll write you all first next time. Promise! Tee hee hee!

Yokwe & Love,

James

***

8/20/10

Dear Mom, Dad, Grandmother, family, friends, and the like,

It's 8/19 for you. This covers 8/20—9/3/10; I'm mailing to you on 9/27/10.

Yokwe yokwe jen Bikarej! Literally, I'm eating my fourth coconut as I write this (today, that is). And wash the fish head off my hands. After eating, showering, and talking (mona, tutu, im bwebuenado) it is now late, so I am going to bed.

***

8/22/10

Bikarej is great so far! It is very quiet and very relaxing. Or at least at times. Almost every day, I have gone to the school and cleaned and organized. Apparently, no one that has "graduated" BES has been to high school in 5 years. The building itself makes you think no one has been in there for 10. And you try to figure this out: grades K/1, 2/3, 4/5/6, and 7/8; 3 classrooms; 4 teachers (one of which does not speak Marshallese); no bathroom or outhouse ... I think I will be teaching 7/8 from 8:30—10, and 4/5/6 from 1—2:30. Everyday. And I don't have the textbooks yet. School starts tomorrow! (Extra tidbit—the school has a $365 budget for the year.) I am a bit nervous to start, especially with the pressure now being put on me. I am trying to use the budget money for 1 of 4 initiatives: 1) outhouse 2) new classrooms (plywood in the middle of an existing room) 3) repaint the school 4) gardening club. Woo!

Enough work, let's talk play. This place is gorgeous. I live on a "rock island" that comes off the side of the main island. I, with my family, live under a big tree. My room is about 10' x 8', and I have to duck to get in. My baba is a landowner/councilman/fisherman; my mama is a housewife. My other baba is a copra maker/fisherman; my other mama is a housewife. Families are a little different here. I am still trying to figure it out. Just popped in my head: I received 3 packages so far (#4 ... and two others). The $1 made it here swimmingly. I may be able to make it to Majuro once or twice a month. (I can go in with my baba to sell fish.) If that is the case, I may be able to get mail and contact y'all. (The boat to Bikarej is broken, but my baba has a boat, cuz he cool like that). School must be started right now for Simone (and Madison!). How is that? And André and John should be in Spain? Maybe? I did get Steven's letter and it made me giggle (tee hee hee!). I had a dream the other night. I was back in B.R. with all y'all—it was weird waking up (thanks to rain in the face) and realizing where I am.

One week down, and a whole lot more to go! Today, two of my brothers and I went for a "jambo" around the island. One is 13, the other 10. They had a machete and gumption to impress—they climbed a coconut tree and got me a few, made a basket of palm fronds that they used to hold some clams they just opened, and showed me a giant "footprint" (I'll explain that part when I learn more Marshallese and hear the accompanying story). This place is pretty spiffy, so I know it will be a good year. Now I need to stop procrastinating and continue my lesson planning.

***

8/27/10

Week one of school complete! I try to lesson plan, but it just does not work out as you plan, you know? My 7/8 class has potential (especially the girls, it seems). My 4/5/6 is tough—20 students, a few with definite learning disabilities, and a wide range of ages. It is pretty tough. Tomorrow, there is a Kemem (1st birthday) for some baby. I am promised a lot of food (and a good time). And on Sunday, I am going to an island on the other side of my lagoon for a day of "Marshallese custom" (manit majel). So this weekend should be pretty hectic for me, with a whole lotta new experiences of this culture! All is very well here; I am trying to get used to 730—9 o'clock bedtime during the workweek. The weekends should be a good change of pace! I practice Marshallese with the kids at home while they practice English (and I am teaching one of my babas Spanish—hectic!). Oh, and you would like to know, I am sure, about one experience: Catholic Church. I went in Majuro a week (?) back. It happened to be on the Assumption (the namesake of the Church). So it was extra special. More "islandy," especially during the procession of the host. A bunch of teenagers did a slow dance and stopped 3 times to do a slow spin move. There were fans, open windows, a whole lot of incense, and a projector that had the words to Marshallese songs (like everybody's favorite "I Yokwe Jesus"). I went with some other WorldTeach people. How is Steven's internship? Any news on New York, or my friend's brother that said that she said she would send an email? How's all them D.s, Heffrons, Morans, Brantons, deBrueys, McCrarys, Cardinales, and such? Peer+Plus, LLC staying busy? Hopefully, there will be more to the letter, then I will send it soon! But know I am thinking of y'all (which, again, you will know when you receive the letter).

***

8/29/10

Howdy, pardners. It is Sunday, so it is slooooow today. I will pray with the members of my family who are Baha'i. Next week, I will go to the Church on island. I was asked to pray yesterday before I went fishing with Tarturin (it was a terrible, terrible prayer). Before he fishes, he prays on an island full of ghosts of dead ancestors. We had to be silent, and could only pray on the island. It had an eerie wind, and an ominous looking old, rusty boat that rested on the sand. Tarturin is a great fisherman. He threw a line with some crab bait in the deep blue, coral filled water. He caught 4 in almost no time at all. Luck ran out, however, when a reef shark grew suspicious of our activity. It was about 10 feet away. It wasn't scary, though. I found myself more curious (which is not a good thing, I'm sure). We went to a small island later, and we drank some coconut and watched the tide rise. (Do not, if you report this to others, put this in an email or mention it outside the letter; this following part.) Later at night, I was playing with my brothers and sisters when my other baba came home from fishing.

They looked at him like a deer in the headlights. Some of the girls ran away. I stayed where I was. He walked into his house and I heard his wife start yelling something, a glass break, I think a thud, him yelling, her yelling something, and silence. One of the boys tried, during part of the altercation, to show me "head, shoulders ..." to either distract me or himself. Domestic abuse is a problem, I have heard, in the Marshalls (just like in many places). I don't know if he hit her, vice versa, or if it was verbal with a thrown glass. It made me upset, but what can I do? My gut instinct was get up and walk in, but my contract says I cannot interfere with any altercation. Tarturin came over to me and told me to follow him, so I did. (End that part.) We went to a Kemem, the 1 year old birthday party! There were a ton of people (maybe 100?), enough food for 300 (including ham, hot dogs, breadfruit, rice, and some Marshallese treats), a lot of songs; and I got to be the guest of honor. They laid out a special mat and gave me a necklace of shells. I got to sit by the mama and the birthday boy. It was really cool. We walked home along the beach with bowls of food (leftovers for breakfast!) and looked at all the stars in the sky. Today, Tarturin gave me a note. He found a message in a bottle a month ago, but he cannot read too well. Plus, the ink had really faded. It took a little while, but I think I got it! Mike and Anna were in Mexico in October '08. They are from New Mexico. It had an address, so I wrote a letter! Tarturin and I signed it. Who knew that messages in a bottle can work? I intend on doing it now, and maybe have the school do one, too. It would make for an awesome pen-pal scenario (probably years later ... if at all). Otherwise, it is quiet, the breeze is nice, and will write more letters, journals, lesson plans, and read! Oh, I think my solar charger got wet and is not working, but it kind of does sometimes. I might be able to charge it at school, or maybe Majuro. Has football season started? How's Madison & Blake? Anything new at home? I should be in Majuro in two weeks, so I can send this letter (if I actually go) and maybe get online to give quick hellos! I guess I'll put it now, too: Happy Birthday, Mom! I love you!! By the time this gets to you, I'm sure it'll be late September.

***

9/3/10

I have decided to make my letters "monthly," so this will conclude my August letter. Hopefully, this letter will be in the mail next week. I am looking forward to going to the "big city," even if it will only be a day or two. I have a lot of letters going out, some to the WT crew, some to friends and such, and the "message in a bottle" couple. Today was a good day at school. It is amazing how short and how long the hour and a half (x2) can go. It's hard working the 4/5/6 with all of its students, ages, and disabilities. In addition, they keep confusing the Marshallese alphabet with the English (so words like "paper" become "baber" or "beber" on tests). I think I will move into a phonics book next. I have a funny story to tell, but it is too much to write. Just remind me to tell you about "Mulmul Day." It went from "cultural experience" to political and educational scandal. Otherwise, all as calm, it's hot, it's got a nice breeze, and a lot of the tiny mosquitoes that are particularly interested in my feet. I should be fishing this weekend off a small island? Next week, Majuro? I am starting to teach (or reeducate) my babas on reading. Target book: Aesop's Fables. So far, they think they are hysterical (after I explain what they just read). Oh, I started a Tuesday/Thursday writing class after school, and will be adding a lesson plan that will be translated into Marshallese (a sort of "living outside Bikarej class that will include budgeting, proper notetaking, how to apply to a job, etc.). If they don't stay outer island, they will likely move to Majuro to work (or go to high school!), so I want to prep them for "urban living." So that is my life here so far! I hope everyone is doing well. I miss all of you a lot, and you know I love yall! I have enjoyed my "monthly letters" in the photo album so far. In the meantime, I hope to hear from yall soon, and I will start on the September letter sometime in the near future. I love you, and I will communicate when I can!

Love and Yokwe,

James

P.S. Thanks for the map, and package from Dre' and John w/ notebooks and jolly ranchers! My baba went into town early this morning, so I didn't know. But thank you so much! I love yall! (And get better, mama!)

***

8/20/10

Dear Grandpa and Aunt D,

It's 8/19 for you!

Yokwe jan Bikarej! I have been on my island for a couple of days. I figure I would update you with some tidbits as the days go on. It is beautiful here—a true tropical island paradise—minus the mosquitos and constant heat … and the terrible education system. I am the first non-Marshallese teacher in over 20 years, and it's been 5 since any student has made it to high school. The last three days, I had gone to school to clean, dust, and organize. I stay hydrated with plenty of ni (coconut). The kindness and generosity of the Marshallese is unsurpassed. Even when I try to help, they tell me to sit; yet they give me a woven mat to sit on. I might be spoiled and fat(ter) when I get back if this keeps up. On Saturday, my baba (Kyotok) and I will go fishing. Before I relax, though, I have a PTA meeting to lead on Friday. I teach two classes for 90 minutes each, every day. Hopefully, I can be a successful teacher and still relax and explore. My Marshallese is slowly improving, so that is good. I've been working really hard - I hope to get at least one student into high school. Lesson plans and school cleanups define my days thus far. To get to school, I walked through eight pretty dense jungle filled with coconut trees (yes, I watch for falling ones), pigs, and chickens. It really is otherworldly.

***

8/25/10

So much work to do! Day 3 (of many) complete, and teaching is slow. These kids do not even know how to take notes, and I cannot explain it to well because they know so little English. The kids are extremely bright, but this environment does not foster academics. Yesterday, two of my little brother and I went for a jambo around the island. It was absolutely fantastic! And they do so much to impress and give over here. For example, one climbed up a coconut tree so one could drink while the other made a basket of palm fronds. The basket was for all of the giant clams they opened with their chetis (Marshallese for machete?). We ate and rested and went around the island. They pointed out the different trees and spiders to me (and they are huge). So far, I am in good health and spirits, and I wish the same for yall. I cannot wait to hear back from yall (once I actually send the letter, of course).

***

8/29/10

Hey, yall! Anthony in two weeks, I can send a letter! Maybe … This weekend was full of adventure! I went fishing with my babas in the crystal clear, deep blue water. Among all of the coral in the lagoon are a <u>ton</u> of fish. In 10 minutes, he caught 4. But not without a price. His bait attracted a reef shark. It was about 10 feet away. I have never (knowingly) been so close to shark. We had to pray at an island full of the ghosts of dead ancestors for a good fishing day. It included digging up crabs and mashing some up. Then, that night, we went to a Kemem (a birthday party for when someone turns one). In the Marshalls, year one is the big deal (it means you are healthy enough to live). We went, and I was greeted by the councilwoman's son with an amimono (a handicraft). It was a necklace of shells. Then they placed a special mat for me by the house, right next to the mother and the birthday boy. Everyone sang songs, said prayers, threw candy, and ate, and ate, and ate. These people eat so much. It is ridiculous. I sat, and ate, and laughed. I had no idea what was going on or what would happen. I would entertain some kids, then I would be given some chicken and rice, then people would throw money in a tub for the baby. It was a beautiful ceremony and a beautiful night. So many stars and shooting stars in the backdrop of the Milky Way, running across the length of the island. It is peaceful and calm here. I find myself often staring at the ocean and losing myself to thoughts or simply the sound of the waves.

***

9/3/10

Well, I think I am online to end the letter right around here. I will let you know that I am on lunch break from school (which is not long enough …) and enjoying the 90.1 degrees F. weather! How's the weather back at home? School is in week 2, and we're trying to keep the pace up. I started a voluntary, after-school writing class, and a.m. Getting ready to lesson plan for a once-a-week "living outside Bikarej" class that will include lessons on budgeting, using a washing machine, STD prevention, notetaking skills, etc. Yesterday, we had no school because of a 0-2 time a year phenomenon where these fish swim close to shore. A traditional trap set made of coconut leaves, and the entire island goes in to help. My sleep has been getting better, and I can have very small, short, simple Marshallese conversations. It's amazing how small the world is here. I have fairly limited in travel (especially on school days—school home). But I am enjoying it, and I am soaking it all up (and all of the sun—it's pretty bright here near the equator). I hope yall are doing well, and know that I think of yall all the time. I love yall, and do miss yall very much. Thank you for having us all for that "last supper" at Beau Chene—it was good to see all of the family before coming out here! Time to walk back to school. I love yall!

I have decided to send letters in this order: Jess, Alongias, Chris, Brett. Can you try to make sure everyone gets the messages? Of course, I will also respond to individual letters! These are just general ones … updates, if you will.

Love,

James

***

8/20/10

Dear Doug, Rachel, Finn, & all,

It's 8/19 there.

Yokwe jan Bikarej! I have been on Bikarej for only a few days, but I have not had time to dilly-dally, as it were. I have been busy helping clean and organize the school. It is a huge mess. It turns out no one from Bikarej has been eligible to go to high school in 5 years, which is really surprising because it seems like no one has opened the doors to the place in 10. Every day I walk through the jungle to go to my school (about 1 mile away) and am often met by chickens, pigs, and children (and today, rain). The principle, Alex, has basically left me in charge … which may not be a good thing. I do have plans for the school—repaint the outside, start a garden, create a mascot (The Fighting Pandanus?), and maybe add another classroom. We start on Monday. I will be teaching grade 7/8 from 830-10, and 4/5/6 from 1-230, everyday. There are only 3 classrooms, so we are trying to figure out where K/1 and 2/3 will go (hence "build" another classroom—put up the plywood wall?).

As my snorkel booties were stolen, I have yet to snorkel in one of my two lagoons, but hopefully soon (when my parents send a pair). My home is pretty cool. I live on a "rock island" that juts out of the side of the atoll, under a big tree (the "loej tree" (sp?)). I have a mama (Imita), a baba (Kyotok), two dogs, five chickens, and a pig or two. My baba is a landowner, since he's a big deal. He's also a fisherman (so a lot of fish for the family, dog). While the boat to the …(?) can go to Majuro more often (about 1.5 hours away) and check my mail (because chances are mail ain't coming here). I will update this letter over time to make it worthwhile. Ak kiio, ijetal nan mon jikuul (but now, I am going to school).

***

8/27/10

I didn't think I would get cold here. The winds can be pretty strong, and in combination with the rain, I get a little chilly. Currently, it is pouring, so I decided to continue the letter. The first week of school is complete! 39 more to go! I have already noticed my "shining stars" in class. It sucks, though, teaching so wide a range of ages (especially 4/5/6—between 8 and 12) and learning capabilities (definitely have some learning issues). I am currently not sure how to meet all the needs so I will just keep going. Next week, I plan on starting a twice-a-week writing class after school, and a weekly "living outside the Marshalls" class (like, for the boys, how to wash clothes, or how to budget money). But it's not all work here, my friends! Oh, no! Yesterday, two of my brothers and I went on an adventure for clams. It was like oyster shucking … for men. These things were gi-normous. You had to shuck with a machete. And, boy, are they different from oysters. A lot more … parts. I almost blew chunks. They made a basket out of fronds and got coconuts and we chilled out ("kakije" in Marshallese) by the lagoon. These kids are my, according to custom, brothers and sisters. I had to dad and two moms. (…) The other night, one of my babas and (…) (I know what I am about to say sounds like it's from a cheesy movie) he asked me "why do Americans carry guns? Why murder? I just stared at him for a second and said "I don't know." He said, "This is why I live here. Marshallese … Simple. Good. No murder." Everyone is so nice and giving, and not just cuz I'm a guest. I am a part of their family. (On a funnier, lighter note, he then told me where I could buy pot on the main island. I started cracking up laughing.)

***

8/29/10

This weekend has been great for me. I hope your weekend has gone well, too! Yesterday, I went fishing with my baba. He is incredible! Literally 4 fish in 10 minutes. Then, luck ran out when his bait began to attract a reef shark. It was about 10 feet away, the closest I've been to a shark (or at least known about it). It was weird, though. I didn't have a "primal fear" reaction. Nor was I excited. It just was. We had to walk past an island of dead ancestors, so I was asked to pray. Unless you are praying, you cannot talk even when you are near the island. If you do, the fishing will be bad. So, I prayed the following: "Dear Lord, let us please catch fish … today … With what we got. Let us catch fish all the time. Amen." A great prayer, I know. He then took me to a secluded island full of coconuts, crabs, and coconut crabs. We drank coconuts and spoke of storms and beards. I cannot wait til I get better at Marshallese so I can speak about things on a deeper level. Speaking slowly and on "surface level " for the last few weeks (both in Majuro and Bikarej) is becoming, honestly, annoying. I know there is more I can say, I just don't have a command of the language. Being relegated to "I ate breadfruit last night. Yum." is just not that stimulating. (…)

*I went to a Kemem (A 1-year old birthday party; a big deal here). And I was a guest of honor. They laid out a special mat by the house, gave me a ton of food, and gave me a beautiful shell necklace (amimono). I got to sit by the mom and the birthday boy. Everyone sang songs, threw candy and dollar bills, and ran around. I just sat there laughing and saying thank you to all of the food and hospitality. The party lasted until midnight, a big deal here in Bikarej (bedtime is around 730-9). Today, Tarturin (one of my babas) retrieved a note he had. Last month, he found a message in a bottle. The note was waaay faded and in English, so he held on to it. It took a while to decipher, but it was so cool. Mike and Anna from New Mexico, 10-21-08. I am going to send them a letter from Tarturin and myself. They sent it from Mexico. It is so amazing: a vodka bottle with a note made it from Mexico to Bikarej. I am going to send one, myself, while I am here. I did it years ago, but who knows? It might lead somewhere crazy! All in all, things move slowly here, so when anything exciting happens, it becomes a **whole** lot more. I read the note about 10 times after it was deciphered for the family, and my letter 3 times. Then we looked at a map to figure out its origins and where I'm from and such. However, I was exposed to a harsher reality last night. I was talking to my kids last night, and their dad (not Tarturin) came home. They looked like prairie dogs, sticking their heads up. Some ran. I got confused. He went into his house … silence … Then I heard a woman started yelling something. Glass broke, he yelled something. I think I heard a thud. An older girl ran in the house. She ran out with a baby. More yelling from the woman. Then it got quiet. I had heard spousal abuse is a big problem here. I wanted to do something. My contract says I have to stay out of all altercations. I don't know if he hit her, or she him, or if it was all verbal. I felt sick, though. The kids kept their faces down. One (after his face was lowered) raised it up and tried to distract me (and himself) by teaching "Head, shoulders …" in Marshallese. Then Tarturin came up to me and told me to follow him to the Kemem. So Paradise is not always what it seems. There are, like anywhere, a lot of issues. If there were not any issues, I would not be here, right?*

***

9/3/10

This will be my last update for the Alongia letter. The next letter will go to Chris. Currently, I just ate some sardines and mulmul (a small fish). I will have to tell you in person about "Mulmul Day"—it was very complicated. I am on lunch break from school where my grade 7/8 moves very quickly; my 4/5/6 is different because of all of the kids/ages/learning disabilities. A cool breeze is rolling on this 90° day ... like it had been (more or less) everyday. Everything is fine over here, so it is all good. I've been thinking of all y'all, and I hope all is well. Doug—Can you still send me some Halloween, Christmas, Mardi Gras stuff? How's Finn? Still at Izzo's? What's new? Rachel—How's school going? And parenting? Luis—How's the internship? Keep it classy? Jess—How's law school? You miss France? Marianne—Still at Chimes? How's the haunt? Kaylee—I didn't poop for three days until today. What's new with you? Katy—Still painting, or are you tailgating? Craig and Adam—new song idea. Remind me in a year. How yall doing? Bobby—You still chazzin'? Cuz you better. Rhett—Get off the computer. Now get on this. Don't worry, I can reach you from here. :-) Ryan—How's not being in school? Karley—Do you have a real major yet?– How's Disney? What character are you safeguarding? And how's the violin? Kate—I will eat wacko. Chris—workin' on that a/c? How's school? Sarah—you behavin'? Give Chris a spanking for me. Brett—how's family and bidness? Oh, and how's Marley!? Keepin' it strong? Ali—You still doing the PF Chang thang? If I missed anyone, I'm sorry! I hope I can hear from all of you, or at least know everyone is good. Go to the haunt for me and swig it down. I might be thousands of miles away on an island 3 miles around, but you know I love all yall. But for realz, I miss yall, and I know yall are doing awesome things. Keep that funk alive.

Laurels and bayleaves,

James

***

8/27/10

Dear Ms. Carolyn,

It's 8/26 there.

Yokwe! I am in Bikarej right now, a part of the Arno Atoll in the Marshall Islands. I wanted to send you a how-do-ya-do and thank you letter, as you helped me get here! Hope all is well and that your family is doing great. I just finished week one school. I have been working a lot. Books were all over the school, dust covered everything, the chalkboards are old and abused, and there is no outhouse. It has been 5 years since someone has even been eligible for high school. I teach two classes: grade 7/8 and grade 4/5/6 (at 830-10 and 1-230 every day, respectively). It is hard working with so wide an age range (coupled with definite disabilities in some). But don't let this sound all bad to you—these kids are bright, and what they lack in academics, they definitely have in an understanding of music, art, and nature. These kids all draw in vivid detail, make mats out of palm fronds, fish, can handle a blade, cook, climb trees, husk coconuts, and know everything about the surroundings. They love to play, but they all work really hard. They are absolutely obedient, from picking up leaves, husking coconuts, to scaling fish, and penning the pigs. I live on a "rock island" that comes off the side of the atoll. My home is under a tree that overlooks the Pacific. The water is crystal clear, and the coral is unreal. (New pen!) Hopefully, I will be able to get some pictures on the internet so everyone can see how pretty it is and how great the people are.

Luckily I am about an hour or so away from Majuro, the main island. So, if there is an emergency or if I want a burger, I'm not too far out. Granted, I cannot go too often (other people's schedules), but we will see! Just today, I went fishing with two of my brothers and one of my babas. Three fish were caught. I also saw bright, bright blue starfish and two reef sharks! To look out at the crystal blue lagoon and see little islands encapsulate you in this little hemisphere is so cool and so breathtaking. It really is a type of beauty I have never experienced (and would not have experienced without your help!). So again, thank you so much! I hope this letter finds you well! Mail takes some time in these parts, so I don't know when I can send this (or when you would receive it). But no worries, and <u>thank you</u> so much again!

Yokwe and Love,

James

***

8/28/10

Dear David,

It's 8/27 there.

Yokwe jan Bikarej! I wanted to write to say kommol tata (or thank you!) for making it possible for me to come to the Marshall Islands. But my letter won't stop there, my friend! I also wanted to let you know about this place. First off, the education system here sucks. And as that is what I'm here to address in World Teach, I should tell you that the school I'm working out has not had one student eligible for high school in five years. The building looked like it had not been open in ten. Before I organized books (scattered everywhere), swept, and took posters down from between the years 1970 - 1992, it looked terrible. I teach two English classes: grade 7/8 (8:30 to 10) and grade 4/5/6 (1 to 230) every day. As far as formal academics, I hope I can help make a difference. I have a few tricks up my sleeve that I hope inspire academic excellence. It should be said, though, these kids are total geniuses in the realms of art, music, and nature. I thought shucking oysters was a task, but opening a clam is a horse of a different color. The Marshallese make it seem so easy. My Marshallese is slowly improving so I am starting to hear songs and old stories. Much to my chagrin, I keep thinking in Spanish (and it's made harder when one of my two "babas" (dads) wants me to teach him Spanish, while my other baba is trying to teach me some Japanese). The coral here is gorgeous, a lot of it almost totally untouched. My home is awesome: a rock island that comes off the side of the island. It is under a very big tree that, when seen from far away, appears to encompass the entire rock island. I'm here with 2 mamas, 2 babas, a lot of kids, chickens, a pig, a dog, and a cat. The Marshallese are extremely generous people, constantly giving coconuts and food to not only me, but to everyone. In some ways, it's almost as egalitarian as you can get. They are the people who, as I like to put it, "use both hands to eat." They eat when they are hungry, scratch when they are itchy, sleep when they are tired. And while it almost seems like they do whatever they want at all times, the Marshallese have a huge respect for one another, constantly helping do chores, make copra, or go fishing. I find myself very happy here. I may have an opportunity to do World Teach in another country after this year. They are, in opening (possibly) new programs in India, Cambodia, and Yap, and they may want Field Directors. So between looking at a job or continuing school, I am still unsure. I will try not to leave you in the dark about the University of Hawai'i. But, speaking of school, how's being back at LSU? And how was Peru? I hope the dig has continued to go well. And I really do want to go—when is the next excavation (obviously, summer '11)? How are classes without dear old James? I hope you and all of the staff are doing well! And I hope you wet your whistle at Chimes often! When I get back to Louisiana, hopefully we can throw back a pint and bwenbwenato (talk and make stories)! In the meantime, I don't know when this letter will be sent, but I hope it finds you well and finds you quickly. And <u>thank you</u> so much again!

Best in all that you do and YOKWE,

James

***

9/8/10

Dear Brandon,

It's 9/7 there.

If you read this to your parents, edit some things out!

Yokwe jan Bikarej! You are too far from Louisiana for my friend updates, so "why not write my ol' chum Brandon," I says to meself, I says. What haddnin, boyeee? Still doing school, or is my boy a big, strong nurse now? And is there a mustache about his fine lip? I'm in the Marshall Islands in the Arno Atoll on the island of Bikarej. It is pretty small and has about 200 or less people. Being a teacher is pretty tough, but very fun. I teach grade 7/8 and 4/5/6 for 90 minutes each, everyday. I also teach an afterschool writing class twice a week. I started a "life outside of Bikarej" class on Fridays, and may start an adult class on Wednesdays. Grade 5/6 is really hard—they range in so many ages, mentalities, and learning disabilities. Some just don't understand while others really don't. But my island is beautiful. I can hear the ocean all the time. The men constantly make copra from coconuts or go fishing (which you would like). Net, line, spear, machete, traditional traps. They do it all. The ocean is so blue, the coral is awesome (but my snorkel shoes were stolen over a month ago on the main island, so I have not dived here), and I eat like a king (which I did not expect). Lobster, fish, clam, crab, coconut crab, turtle, coconut, breadfruit, and pandanus. I mix my "hanging out" with my little brothers and sisters, my students, and a bunch of men ranging in ages of 19 to 65. I have gone all over the island on jambos (walkabouts). The jungle is so awesome. It's so lush, the browns and greens clashing in such a dark area. The canopy blocks out a lot of sun, making all of the lizards and spiders well hidden. Which is scary, cuz they are huge. Everyone is very kind here, giving constantly and singing songs to me at different times. I was the "guest of honor" at a kemem, a one year old birthday party extravaganza. I got to sit next to the baby, which as we all know, I just love to do at all parties. Sit next to the baby. I got this necklace. So … have fun imagining that, big boy. I am growing my beard, too, Mr. Man. Hopefully we can meet in a year …touch beards.

***

9/17/10

It is 8 in the morning. School is getting ready to start. The last two weeks I was supposed to "go into town" but that has not come to fruition. "Next week" they keep saying. So what else can I do but wait? I went clamming the other day, but came up empty-handed. My brothers and sisters are little pros. You would love all the fishing, my blonde chum. These people work all the time, and always relax. I think the only part you wouldn't like is the whole "alcohol is illegal" part. It's not bad. I don't miss too much like t.v. (never had one really) or certain foods (although a burger would be nice) or even liquor (although a beer with the burger would be nice). I really miss the conversation and the chums I would have that with. At the same time, I think about this stuff the most when I write letters. I get reminiscent at times, sure. But mostly I enjoy myself and appreciate how lucky I have been to get an opportunity like this. And how is football season? You gone to any games or tailgate yet? How's LSU doing? And the Saints? And how's yo mama an' dem? I hope all is well on your end. Right now, I look forward to "next week," Christmas, and some other things that happen around the island. I also love teaching—I could definitely see myself continuing in some similar vein (if not here, somewhere else). Of course, I still have a lot of time to think and decide on such "life decisions." But until then, I will enjoy this gorgeous weather, plentiful coconuts, and clear blue water. Now, I am about to teach. Yokwe for now!

***

9/24/10

I made it to the big city of Majuro. I don't know what time is. I got in last night and ate his salad and had a beer. So good. I should get on the net a little. So I hope I can talk to you on it (of course, you will not know this sentiment until next week). (Or a few weeks? Who knows). One thing I've noticed and I wonder if you have since you live in rural environs, is how loud the city is. It hit me for a second yesterday. It's been so quiet the last few weeks. Being a restaurant with like a madhouse. It was weird. I am at a dorm for the high school teachers, so I am with my World Teach buddies right now. It is nice to talk with them and share stories. We stayed up til midnight (like 4 hours past our bedtime) just relaying all of our teaching horror stories and some of the gossip (tee hee!). I'll fill you in on a little thing that the kids don't know back in home: (…) Don't show this to your parents. Please. Haha. There are no ladies really on my island (which is good). How is all your stuff back home? How are your parents? Eat some cheese potatoes for me. I want to go to your camp, drinks bottles of Yoohoo, to have a ball, talkin bout sitchyations and beards. How your bros? And your nephew? Are there more than one? How's the CAN (?) program—nursing? You better finish on top. I love you, man, and I hope this letter finds you and your family well. Be good, little bum fluff.

Sincerely, Cordially, Love,

James

***

9/8/10

Dear Michelle, Patrick, Madison, Blake, and all family and friends,

It's 9/7 there.

What up in da hizzouse? I decided to write the "family" letter to you this month, McCrary clan. But I'll keep it personal on the get go. How yall doin? Tell Madison and Blake that I love them and that I miss them a ton! Of course, I miss yall, too, Michelle and Patrick. Yall tailgating? How's the Country Club? And how is school? Is Blake at Seven Oaks? And how is Madison handling another year at SEAS? I want to hear all about it. I just finished teaching and am currently writing in the office. It just finished raining (and it has been doing all day, and is prone to do all the time), so there is a really nice breeze blowing. It's 86°, so it's pretty cold. The weather over there, I am hoping, is starting to cool? Maybe a smidge? I had a meeting with the principal today. We are trying to still figure out if we should build a bathroom or another "classroom." School is going fairly smoothly. If all goes according to plan, my 4/5/6 class will turn into a 5/6 class ... but knows if/when. Island life is ... tough yet surprisingly easy. Truly, everyday is the same. At the same time, I learn and experience all new things. My conversation skills are centered around food, sleep, and some school things; other than fishing, I wonder what everyone talks about. I am picking it up slowly, but I am learning. In three months, I'll be a friggin machine. I have eaten a ton here (really, they keep feeding me and themselves). The other night, some of the fathers on the island brought me baskets and baskets of food, sang songs, and then made a speech thanking me for coming. I was told me this would happen about five hours beforehand—and that I should sing songs and prepare a speech. Awesome. "One Day" by Matisyahu, "Aiko Aiko," and "They All Ax For You." I thanked them for all of their kindness, and told them to also thank the other teachers because they are working hard as well. I have eaten turtle (which was gruesomely terrible and interesting to watch, but so good), coconut crab, all sorts of fish, and lobster. When I tell you big ... I mean this thing was huge (the lobster). Easily the biggest I've seen. I had eaten lobster 3 times now, too. It's crazy. One of my dads (baba) is already planning my birthday party. He says he will give me a "likatu" (a good looking girl) because that is "Marshallese custom" (manit majel). I called him "bwebwe" (crazy or an idiot). During the week, I am working a lot—wake at 6:30, walk to school at 7:30, start at 8:30 and generally stay til 4 to 5. On the weekend, I am "ri-majel" (a Marshallese person) - they finally let me start "working." I went to church, and am opening a few coconuts. It will start building up, I hope, to where I can help make copra and go fishing (which should start officially next week). Oh, and they start practicing soon for Christmas stuff. I won't give too much away, but apparently all the islands in Arno all go to one island (called Arno), so it is supposed to be huge. I know I've got a few months, but when there is nothing else, you find you really, really, really look forward to things.

✳✳✳

9/11/10

Major blow today. After a week and a half of saying I was going to Majuro, we are not going this week. I should have known not to expect things in this place. In a world of "awa in mujel" (Marshallese time), I should remember to not necessarily "keep my hopes up" about certain things. One more week, they said. Besides being able to send my mail and maybe get online, I wanted to go to the Ministry of Education (MOE) and get a school calendar. We don't have one. And an alarm clock. Mine fell in the water. Of course, I also fell in the water with it when I fell out of a boat the other day; I got some of my school things wet, too. Sigh. But things aren't bad here, I am now just in a mood. All week, people were talking about me going to Majuro. Things get built up here. Anywho, we may watch a movie tonight (may), so that could be fun! The tide has been uber high, lately. It is now later in the day. I went on a surprise fishing trip. They are training me on net fishing. When the tide is low, there are a lot of fish to be had. I'm still getting used to eating the fish head. It's not bad. Just complicated. As far as Englishing the people's brains, I've got some kids now phrasing sentences for things they want from me ("May I please take a picture?" "May I please listen to music on your Ipod?" "Will you please make the sound of a cow?") They learn really fast(at least for things they want). Oh, and my 7/8 class took their first test—no Fs! So good sign so far. This week will be the halfway point of the quarter—the 4th week. It is so nuts how fast it's going, and yet it feels like I've been here forever.

***

9/17/10

No Majuro this week either, but it's ok. You will get the August and September letters around the same time. I am <u>definitely</u> going in 7 days—I have been invited to speak at the Councilman's meeting on behalf of Bikarej Elementary to try to garner some things (like plywood for a "new classroom," paint, and (fingers crossed) AT LEAST a half-court for basketball). There is, supposedly, a new "schoolhouse" and a carpenter on a boat in Majuro that is waiting to set off to Bikarej. With a new building and a plywood room, we could have 5 classes at once! I got to go help sell copra the other day on a small island where only one man lives (he runs the store). 100 pounds of copra fetches 413. We had ten bags. The whole process is so long to make copra, but I guess it can be viewed as solely profit (everything in nature supplies the coconuts, they just need knives and ovens). However, the business seems a little shady. Each bag is about 130 lbs., and they lug all these over. They don't get cash in hand (at least, not what I saw). It is kept in a ledger, and the goods you buy are sold by the company that sells the copra. I don't know if it is true, but my baba bought a big container of "biscuits" that cost $50. These things are cheap crackers, $5, $10 in the States. The goods you get are from the company itself, with no cash shown? Hmm … In other news, ants got into the tins. Crafty buggers.

***

9/24/10

Big City livin' in the Majuro!—hopped on a boar and we went over to Ulien to drop someone off. Last minute we picked up my friend, Brooke from Ulien (she had to run out of class and had a big smile on her face). Needless to say, our conversation was non-stop. We grabbed a (?) and ice cream for lunch and headed for NTA for some internet time. I got to update yall a little bit there. I got all the mail and packages. On letter was partially eaten (one from André). Everything is really, really appreciated. The food, the candy (which I will give to the kids), the bag, and the snorkel booties. Biggest thank you is the notebooks! I can give those to my 4/5/6 as "English only" notebooks, which they really need. I had a salad that night with my colleagues and friends. And a beer. No, they *don't go, but the salad was much needed. A primarily card diet, just one month, starts to drag on one's spiritual and physical health.*

* * *

9/25/10

Another friend from Arno arrived yesterday. So we went out, and I had a slice of pepperoni pizza and some sashimi. I looked at Madison and Blake—they're lookin' great! Majuro really is a different place, and I like that I may have the occasional option to jump over here. Plus, getting resources or talking face-to-face with the bosses or MOE people can get more things accomplished or seemingly so. So I missed all of you but Dad on skype. I will try again today. I took a shower (as opposed to a bucket shower), and while there is seemingly one option in the Marshalls (cold), it felt fantastic! And flushing a toilet without having to get water from the ocean with a bucket—what paradises I am experiencing in these urban environs! It is not that strange to be in Majuro, and I know when I get back to Bikarej it won't feel strange either. But it is so weird how drastically different the lifestyles and cultures are from Majuro to Arno. The people on the outer islands definitely seem happier. They had a murder in Majuro about a week ago, which is unheard of here. There are "gangs" starting to form, and alcoholism is on the rise. Outer islands they are more worried about getting food, so tensions are 99% of the time low. Tell André and John I found <u>Pandora's Seed</u>. And it has a lot of good and interesting points. I can relate some of them to this place. All in all, everything is good! I will end the September letter now, and yall should get the August letter around the same time. I love yall a bunch, and thanks for the letters and packages and everything! Oh, and thank you for helping me with the basketball fund! I hope we can do it! I'll try to get more info on it. I love yall and miss yall. Geaux Tigers, and how bout them Saints?

Love,

James

* * *

9/8/10

Dear Chris, Sarah, Vizzinis, and all you chirren,

It's 9/7 there.

This is the next "friend" letter! But, since this letter is addressed to Chris, I'll talk to you first. How's everything, my amigo? How's school? Dude, I thought about how you, me, and Brett used to wrestle as kids and about Aviation Challenge and all kinds of crazy stuff the other day. This place really is good at making one reminiscent about everything. I would love to grab a "meet me at Mellow" beer or watch some Always Sunny or King of the Hill. For now, it'll have to wait. But you should grab a beer and give Sarah a lil 'pankin right on the bum, for me. Life here is crazy and dull. I am in the school office, listening to the waves of the ocean at high tide. While that is calming, I have been stressing over lesson plans for not just grades 7/8 and 4/5/6, but my new writing class after school and my new Friday "life" class (how to live outside of Bikarej, essentially). I also may be starting a once a week adult class. More details to come. The little kids here are great! I love my students.

On the weekends, I hang out with my little sibs and students. On Sunday, I went to Church (and got a wot, or a flower wreath for my head) and ate turtle. Turtle is so delicious; watching it writhe for hours and hours … not so much. Blood and guts and entrails and eggs and more blood. I'm glad turtles don't make sounds. It was tasty, though. And it fed 30 people. Huge things, I tell ya. I'm getting used to eating fish heads … roly-poly fish heads. The eye ain't too shabby. I love the jungles here. They are so thick and lush. There are a ton of huge spiders (sorry, Chris) and lizards. A bunch of old coconuts grow vibrant green moss. The trees provide almost all the food you could need—coconuts for vitamins, breadfruit for carbs, pandanus for dessert. They brought me (my little brothers) to a spot in the jungle that is a mini-swamp called Labiroro—so deep, no one has touched the bottom (or lived to tell about it). It was like a tropical Louisiana. They had trees with almost "cypress knee" roots in a world full of coconut trees and mud so thick it felt like I couldn't move. There was bedrock underneath of old, dead, hard coral. Step in wrong places, you could pierce your sandal. They say there are a ton of fish there, and they will take me fishing. I am reading 3 books right now, yet I never find time to read. Yet I have <u>so much</u> free time. Soon, I hope to have a lightbulb at the house. Once that happens, I can read and write like one as "mad as hatters," to quote Thomas Hardy (I'm also reading poetry!).

***

9/11/10

I was supposed to go to Majuro yesterday. Now it is today. I woke up at 6:30 (per usual) excited and ready to go. About two hours later, my baba shot me down. He said he needed to go fishing. It's fair enough, but it sucks. Everyone built up during the last week I was going into town, so I was genuinely excited and ready. So … I'll send the August letter next week. It's not uncommon to have things scuttle in my room: roaches, crabs, lizards, spiders (big ones—sorry Chris). Last night, I almost 'ad a 'art attack. I heard some serious rustling. I woke up and grabbed my flashlight and saw a tail. And I mean a big thing. I shot up and looked around, and out of the dark this thing pounces. Cat. Friggin cat. I went on a little fishing trip today: learning the art of net fishing. Soon, I hope my training is complete and I can really get into the fishing (especially the spear stuff). My 7/8 graders took their first test, and no one failed (more or less), so I am very proud of them. I have to write tests on the board (which is not really good). There is a copy machine here (I know, right?), but the "big battery" is dead. It has sat idle for its third year now. I wonder if it'd even work. The tide was really high yesterday and the day before. The water was about four feet away from the door. It's nuts, I tell ya. Just nuts. I'm gonna go take a walk to a small island right now and read/stare at the water. Peace out, churren.

***

9/17/10

I will go to Majuro next week, so you should get the September letter and the August letter around the same time. It's Friday, and I have one more class to teach this afternoon. I am sweating a lot from playing their version of soccer. A lot of picking the ball up and just kicking it far. I have been doing some more work now, "moving up in the ranks" so that I can help more with husking and cutting coconut. Cutting copra is pretty tough, but I hope to have it down pat in a month or two (like shucking an oyster). Chris and Brett: I had a dream Mr. Ray (our old bus driver) went Twisted Metal. Kaylee: I had a dream you and I were ambushed by thugs (one of which had an explosive). My dreams here have been nuts-o-cuckoo. My sleep pattern is very strange as well. But the dreams I've had are vivid. Enough so, apparently, that my baba told me I yelled out one night. He told me demons don't live in Bikarej anymore, but they can attack in dreams. He prayed for me, so the demons will leave me alone and I can sleep. I helped him write a letter the other day to his daughter in Majuro. He said he wanted to practice his English, so I figured why not have two people practice? He is very excited; I told him that if the "message in a bottle" people write back, he can write the letter. To all who are in school: how is it going? Dubus should be graduating this semester, so congrats, keep trucking. And you other kids, I got yo backs. If I am in Majuro next week, I will send this letter, too. Who knows when I will go back, ya dig? I can't believe one month of school is finished—it's flying by so fast. This year will be over before I know it. How's Finn and Marley? Sorry my letters are so disjointed!

***

9/24/10

I am in Majuro (as some of you know by the time this letter arrives). I hopped on a boat getting ready to leave. The council-meeting was canceled that I was supposed to go to (pushed back another week). So I had to join a random small fishing crew. I had a burger and ice cream yesterday, and a salad for dinner. SO AMAZING. I am at the dorm at the high school until Sunday. So I hope I can talk to some of you kids on the Facebook. I am with some of the World Teach crew, so it is nice to be with friends. We have all shared our stories (lasting until midnight, which was way past my bedtime of the usual 8-9 o'clock). It was not weird coming into the big city. But I did notice a lot. The main things: how many people there are and how loud it is. The restaurant I went to had more people in it than the total of my school. Everyone talking at once with forks and knives hitting plates—it is amazing how much volume it creates. It was wonderful to be with my WT friends and dish, like I was a school girl again. It makes me realize how truly different everyone's experience is here. I have one of 32 different stories. Teaching in Majuro is completely different (apparently, some 8th graders at one of the schools have started "sex gangs" where the number on your shirt represents how many people you have slept with from the other sex gang—8th graders …). Today is Manit Day (Culture Day), which includes things from a silent auction to breakdancing, along with more traditional activities. It should make for an interesting day. Also, it is still in the works. (I talked to my Mom about it yesterday.) I am trying to raise money back in the states to build a basketball court (even just a half court). I really could use everyone's help, even if it is just a few bucks. Tomorrow, I am going to go around to some of the businesses and ask about prices for a court and a half court. Next time I can, I will update you on the costs. It's the one thing the kids and adults talk about on the island, and if a bunch of us Louisianians could help it would be fantastic! Call my mom (225 388 9822); I am hoping she can collect the money and put it all in a single check. Thank you if you can help! I just read mail—thank you Doug, Rachel, and John! Responses will be slow, but things will come (I swear!). Believe me: letters help so much. And Doug, thanks for letting me know of the All Hallows Eve goodies you will send. Hopefully I can get them on time. I think about you cats everyday, and I hope this letter (and the August letter) reach you well and in a speedy fashion. I love all you guys.

Never,

James

***

10/16/10

Dear D.s and Heffrons,

Yokwe! How are you fair haired chirren doin? And to my aunt, how's my favorite Godmother? I am writing to you on the breezy October day to let you know that all is well in the Pacific. I am taking a kakije (a rest) from my morning jambo (aimless wandering or trip) that took me around most of the island. After working up a big sweat, I ate "jop im ma ippan corned beef" (breadfruit soup with corned beef) with my boss, Annie, who came to do a site visit to observe me teach and ensure I am in good spirits and conditions. Apparently I got all As, so yahoo! I finished the first quarter report cards. There were some good grades, but I also had to dish out some Fs. It's hard to learn (academics) when your other teachers just write on the chalkboard definitions in the back of the book and then walk out of the class room to drink coffee with their friends. Or just giving them another recess. But we're working on it (as much as there is time for, anyway).

I am eating plenty of coconut almost everyday (4 yesterday!) and am trying to keep healthy. Thankfully, they have a pretty localized diet. I've had spam only twice now (which, apparently, is a record for outer islanders). My island is really big, but there has to be less than 200 people on it. There are a ton of plants, and I am going to help my baba do some NGO work. He is in charge of the Jabod-wad ("tip of the coral") district. He has to plant more breadfruit, banana, and papaya (ma, pinana, im kinapu) in an attempt to keep the diet more localized and simultaneously create a small produce market to get some money onto the island. I am trying to be as involved with anything and everything.

Love,

James

***

10/5/10

Dear Ma, Pops, family, friends, and enemies,

I just finished my meal of canned mackerel, rice, and breadfruit. I will never like canned mackerel, I have decided. We almost cancelled school today because of rain, and I and two other students were the only people there. I rang the bell, and within 15 minutes, almost everyone showed up. So far, we have had school on all the proper days, which is a rarity in the RMI. Supposedly, I am going to Majuro in two weeks for a council meeting (the very same one mentioned in the last letter). It has been postponed for over a month. I do not want to miss school, but the principal and my baba (who is a councilman) said I should go (for the basketball court, essentially). Things seem to be moving with forward progress on all counts: we got the parts to our new school room, they will build an outhouse, they are going to build our "new" classroom by walling in with plywood, and they will bring in a big battery that will help charge the copy machine at our school (which I really hope will work). We had people from the Ministry of Education come "train" our new teachers, so they all radioed our requests for us (no paperwork!). As far as school improvement, things seem to be a-ok (but when will the computer and these supplies come, if they come at all?). I am working on a lot of phonics and alphabet stuff with my classes. I have separated "problem letters" into 3 categories: B,P,V (all pronounced "B" to Marshallese), C,G,J,Z (pronounced "shee," basically), and DT (both said "D"). I feel if they can actually separate both letter and sound, their spelling and comprehension will improve (they get "paper," but spell it "beba"). Next quarter, I start teaching the little chirren maybe 3-5 times a week. My work load ain't too shabby, and they have me as their primary English learning took (followed in second by DVDs). It's sad, and hard for them to learn English when their teachers barely know the ABCs (and its pronunciation). And really, their educational knowledge is also on the lower rung of the ladder—I sat in a math class yesterday to watch the teacher and student solve a single math problem together for 30 minutes (167 + (-196) / 205 + (-102)). It made me wince a little bit. But things are not all bad—students are participating and trying more in school. Their self-confidence is rising as far as academics is concerned. So baby steps. Otherwise, perfection in this place. After school today, I was given a big piece of BBQ chicken and rice (I started the letter earlier today) and walked slowly through the jungle. I went on a picnic with my family on Sunday on a small island nearby called Eneca; Beach baseball, sleeping on coconut leaves, and earing pork—not a bad day at all. This weekend, I am supposed to go to Arno (the atoll's namesake) for some fishing and snorkeling (thanks mom and dad!). Next weekend, Majuro. Oh, and it turns out I am the Church's Christmas choir director, specializing in English songs. So I am always busy and yet doing nothing. Also, I will officially teach the entire school English starting October 18. I know it'll be hell, but I consider myself a Dante of sorts. <u>Not</u> really ... it'll be really hard.

***

10/9/10

So I went diving, right? It was pretty nice. I went with my brother, Biku (like "B.Q."), and my student Junior. Right off the bat, when I was putting on my flippers, they yell "shark!" I look and it's this tiny thing, but a shark. It swam past when Junior jumped in with a knife to try to kill it as fast a lightening. A nice little start. The coral is, honestly, decent (we stayed relatively close to the shore just because we had only one knife). A lot of the coral is dead, but I have seen some beautiful coral in the ocean. Anywho, I'm in the water and I hear something. It sounds like yelling. My head emerges from the water and I hear "Shay-mez (my name)—SHARK!" I put my head back in the water and not even five feet away is this shark head. My head came back out of the water and I watched the fin bank right, but a turned around and swam back to our little "base." I'd give the shark maybe 3-5 feet long. Like when I saw a shark when I went fishing, I was not afraid, but more curious. It's my type, however, that probably end up eaten. Anywho, when I head into Majuro next (maybe this week?) then I will try to email and facebook some photos to yall. And I may have some Christmas ideas for yall, and wills send some stuff in the mail. Hopefully I'll hear it soon, but I hope all is well currently at home! We watched Princess and the Frog last night, and I thought of home (it takes place in N.O.). Things are going swimmingly (hyuck, hyuck), and I will update again soon!

***

10/16/10

I might send this letter early. Annie, my boss, is on a site visit, and she may be able to take my mail. So I will regale you with information, happenings, and feelings! My teaching was inspected, and I am not fired. So that's pretty spiffy! Having Annie here has allowed me also to have a lot of English conversation. And unexpected events. Last night, while giving her a welcome party, they were singing songs. Then, my mama began to throw packets of ramen to everyone. The kids hounded on it like a defenseless piñata. In Marshallese custom, you have to sing songs back to those who sing for you. So we sang "You Are My Sunshine" and "Stand By Me" and then we had a big ukulele song fest until about 11. I showed Annie around the island, and seems to be enjoying herself. That meeting in Majuro was cancelled. So as you will see when you get this letter I did not go yet. It may be next week. Look at what I work with here. It's been moved back for 2 months. But my patience grows. I realized I've never given you the names of my family—they know all of yours from the album (thanks for the prayers, ma). So here is a list:

<u>Kyotok and Imita</u>		<u>Tarturin and Jujubean</u>	
Salem (Jeral)	(11)	Frank (Buliju)	(12)
Taji	(5)	Katije	(11)
Taki	(2)	David (Biku)	(8)
Wiskey	(1)	Clora	(4)
Romita	(21)	Billa	(6)

Rod (9) and Rickji (11) are "cousins," but culturally are still my brothers. These are also my relatives just at home. I am related to some "higher ups" and influential people (and as I found out, one of my fellow WT friends, Brooke. She lives with Kyotok's son on an island called Ulien. I'm her uncle!) Hopefully I can get into Majuro by November so I can send you a package of goodies. I'll also try to send some amimonos (handicrafts) for Christmas distribution and gifts. First quarter was officially ended yesterday. It's hard to give Fs to kids … However, it doesn't seem to matter—they still move you to the next grade if you fail (heck, you don't even have to finish the year apparently). I got the package from dad and packages from Mrs. Shackleton (TONY'S!) and the letter from Mrs. Price/Trahant. Thank yall so much! Letters really keep this old soul rolling. It makes me happy to hear the updates and the various brouhahas that are happening back at home. I think I may go swimming soon. How's McCrarys and Cardinales? I'll be writing André and John next month. I don't have an address for Steven, but he should get one by next month, as well as my darling little sister in the same time frame. How's his internship? Is he still ugly? And how is Simone's collegial journey? Is she going to class everyday? Parents: I know yall behavin' in the empty nest. But Grandmother is probably causin' trouble. Keep it up, Grandmother. Tell everyone that I love them and that I'm trying to churn out letters. It is strangely busy, especially right now! I love yall!

10/31/10

Dear André & John,

Yokwe! From the other side of the planet, it's your bro/in-law, James! Oh, and 'hola!' How is Espana? Things here on Bikarej are going great. I just made a coconut jack-o-lantern. I have introduced the concept of Halloween to the kids; but when I put up some decorations, my host mama grew suspicious. There are some spirits and demons on Bikarej. I may be instigating a little ... but there is candy! So, it seems to rest a little easier on people's minds. Besides, we "celebrated" Friday in school. So ... all gravy. How is Madrid? And teaching? Are yall in high schools? I just started teaching kindergarten this quarter. I'm definitely happy I don't have kids right now. Rambunctious rapscallions! All in all, school is good. Just trying to stay creative. I have been teaching a little Spanish on the side. A lot of my kids say "adios" when I leave. One of my friends here calls me "senorita," and I him. Pero, you necesito practicar mi espanol. Mi Marshallese esta una mezcla de espanol y Marshallese, y mi espanol vice versa. Ak, ij katak kajn majel aolep raan, im jab melo klok kajin Jipain wot. Im elukkun emman. (But, I am studying Marshallese everyday and I have not totally forgotten Spanish. So that is really good.) Anywho, my island is pretty awesome. It's really big—so I have less chance of feeling "confined." I have a small lagoon with 6 walkable locations (at low tide); and Bikarej actually has differing landscapes, including a swamp called "Labiroro." It is cooky—it has thick mud on top of an old coral bed. At a certain point, there is a drop. Some say it's so deep ...you can never touch the bottom. I've been learning some stories of "etto im etto" (long, long ago), as well as many songs that are quite catchy. Anthropologically, it's neat to translate new songs and old songs. I like the old songs more, but there are more "western timed" songs nowadays. Those are, for me, easier to learn. Oh, I finished Pandora's Seed. Good book—I agree with many of his findings, but he does go a little more extreme (I think to sell a few more copies ...). But it is good to read that kind of book here. Marshallese, among many Micronesians and Polynesians, have high blood-sugar levels thanks to many western influences. Hence, many of my meals (while differing greatly!) are fried, canned, extra-salted, and served with ultra-sugared tea/coffee/kool-aid. Many people here think diabetes was caused by nuclear bomb testing in 1954 on Bikini Atoll. Funnily enough, diabetes in Marshallese is "tonal" (?) (tawng-al; this word means "sweet," "sugar," "candy" as well). It'll be interesting to talk to yall and hear about your adventures and my adventures. How long are yall gonna be there? It may be possible for some running with the bulls ...? I'll have to definitely see my monetary situation. I'm gonna ge a po', po' man. Thank yall so much, too, for all of the packages! The notebooks are "English-only" notebooks. I'll have to explain the supplies/materials situation in person. It's ridiculous how things work here. But jidik, jidik (little by little). I will continue this letter more later. It's time for singing practice. My little sister's kemem is next weekend (it got cancelled yesterday), and we are practicing Marshallese and English songs. In English—"I'm on Top of the World" and (of course) "The Star Spangled Banner." Oh, a kemem is a one-year old's birthday. Culturally, it's the biggest thing besides a funeral. Big party, lots of food, and a lot of other stuff. It was rare in the past to live healthily in the Marshalls. The one year celebration basically means "you survived, so you'll probably make it here! Wahoo!" So, I better get to singing! Huzzah!

***

11/4/10

Tomorrow will be the last day of a big project I have to do to get TEFL certification. This week has been really busy, full of work and social/cultural stuff. I think I may be able to get some mail off tomorrow - but we'll see. Is there a Spanish equivalent of an apple for a teacher? Like … paella? I get coconuts, which is way more awesome than an apple. Though, the other day when my boss came to observe my teaching, she did bring apples. So delicious. When your diet is relatively the same, it's pretty awesome to shake it up. Best apple I've ever had. Like, I savored it. She brought six, so I cut them up and shared. <u>Everything</u> here is shared. It's funny how I didn't prepare myself for the truest sense of this word. Ipod, camera, flashlight—sure, I knew I would. Mouthwash, sunscreen, my belt … not so much. They just want the things I have, and I hope I don't come off as a douche when I write that. Even the few things I brought to the Marshalls—my room is like a mystical palace. My room may have more things in it than some of these houses. So, my little siblings love to come in and explore all the goodies I possess, and for the most part I share. It's really trying on my cultural upbringing. Not the sharing, but the constant proximity of everybody. The family sleeps on the floor (on "jaki"s, a woven mat), and I tell you they are on top of each other—heads as leg rests, legs as head rests. I've slept with my brothers a few times. I feel a leg thud on my stomach. I look up, my brother has his leg on me, his hand on another's face, and is spooning yet another. And they don't wake up from it. I'm a pansy. But it is so weird not having privacy. I don't think about it often, but there are times when I realize that even when I shower, half of me is still exposed. The good half, so don't worry (you can take that however you want, John ;-)). Things are great, life is wonderful, teaching (English) is tough but I am noticing some progress. I hope all is well in Madrid. That would be so amazing to go back again, but getting to see yall. Hopefully at Christmas, I can either Skype or Facebook or whatever and chat. Know that I love yall, and that I know yall behavin'. Let me know how yall is, if yall are going to stay in Spain, and if there are teaching positions available …? What is the next step for us all? Zoinks!

Love, Yokwe, and Amor,

James

***

11/2/10

Dearest friends and family,

Howdy, howdy! It's James! Yay! It is November 1st back at home. Can you believe it? It's definitely coo-coo. As I'm trying to get TEFL certification, this is my busiest week. Tomorrow, I will be filming my class and assessing myself. Then I have to reteach the lesson and film it, again assessing myself. Plus, I need to make potato salad. I know, right? My sister's, Wojke, Kemmen was cancelled last week (surprised?—no). This week, I was put in charge of potato salad for the kemem that is now this Saturday. At this festivity, I am also to sing songs in Marshallese and English with my baba. Don't worry—we have been practicing. Songs in Kajin belle (English) = "Top of the World" and "The Star Spangled Banner." It should be a good party. Oh, and I never told you (I think …) I am also working with the Church here to teach two Christmas songs in English. "Joy to the World" and "I Saw Three Ships" (my choices) will be performed on Arno (the island) on Christmas. I will be leading the song on stage. So … neat. Haha! We practice every Friday from 8 PM til midnight (They are also singing two Marshallese songs and are doing a dance from Fiji—it's fantastic). So days get busy, busy. Up at 6 a.m. (sometimes 6:45), school from 830-330; Monday radio meeting with World Teach at 4;Tu & Th class with students (writing class) until 4; W & F teacher class—I'm teaching English to the teachers (but Wednesday is for all adults, Friday solely teachers) from 330-430. Yes, I am a ninja. This letter should be received after, but I am sending a package, Ma & Pa, for Christmas stuff. Do with it as you will. Things are going well here. I hear there is mail waiting for me in Majuro, so thanks in advance. Everything you've sent has been handy! Like the candy! We had Halloween on Friday (the first on Bikarej!). I carved a coconut. So … it was spooky. I hope all is well back at home. I love yall lots and miss yall. Ma—don't plan just yet for my return. I'll talk to you in December. I don't know how my return flight and such works. Expect (hopefully received?) a package with goodies! Until I update, ciao!

***

11/7/10

Well, the package of goodies was not sent yet, but hopefully soon they'll be in the mail. The kemem was postponed again. It might be today. It's been pouring down rain a lot lately, which has caused a lot of things to be moved or postponed. The carpenters, for example, (and not the band) still have yet to come build the new school building. They were "officially" supposed to come last week. Now it (the answer) has changed to "supposed to be sometime this month." So my classroom is still full of 2x4s, cement, and paint. My baba returned with the supplies for potato salad (these are the kind of things my stipend are used for, haha) and he also brought me mail! I got a marvelous package from the Schanevilles, the one yall sent with Michelle's stuff, a Halloween (better late than never) package from Doug, letters from some of my WR counterparts, and a letter from Mike and Anna Murphy in Albuquerque, NM!* (*In the left margin of the letter was this note: Oh, and Kevin's letter! That kid is a good writer! Tell him and the Morans I say hello!) So it definitely was an exciting batch! Really everything yall send is appreciated, and I hope I can get my package off sooner than later. It was cool to see how excited Tarturin got about the letter from the Murphys. He is nervous, but I told him he should write them back. Lately, I've been practicing English and Marshallese songs with my baba Tarturin, and some friends Jonwi and Mon Kuk (the latter's name means "kitchen," which everyone loves to joke about). I won't lie, we don't sound too shabby. And I'm picking up some songs and new vocab. The trade is good, since they also get some English songs. They have one AM station here that plays some American music, some Marshallese, and has BBC updates. Their knowledge of some English music is relegated to just humming or mumbling. Learning the words, they are like, "ooo! Melele!" (I understand!) I will see what I can do for today to help with the kemem. All the kids are in and out of my room, looking at the books, or playing UNO. So I don't know how much works to be done. Next update—hopefully we can celebrate a one year old's birthday!

***

11/13/10

It doesn't matter if it's in a tin, ants can get to it. Haha! But, ants are protein, right? The kemem was fun. I took a lot of pictures (until my camera pooped out). I tried to help—I got to season some food, like the sashimi and badikdik (sp?—pigs blood). I helped a little with the pig prep. I also hung up a light (yay 6'3"!). Mostly I was photographer and singer! Our band, The Beatles, made of my baba Tarturin, Mon Kuk, Tandin, and myself, sang quite an array of songs. It's hard to be a smash hit when people are more concerned about the candy being hurled everywhere, or when mama sprayed perfume on all of us and made me sing into a chicken leg like a microphone. (I'll have to explain all of this later …). But it was a good time—and on a school night. Very tiring. Iton Kwainini im jambo nan Labiroro. Kiio, imij jeje, ak kwe jela ij yokwe eok! (Oh, and the package is in Majuro, so it should be sent soon, I hope!).

***

11/18/10

Again about the package: I superseded your shell request! Man, I am good. So now I have a cool small box. I hope the package gets to you soon. Fairly certain it was sent. This will be the end of my update for this month. **I should name these something … "Jambos with James."** *Things are good. Just a "day" today. Taught K-8 today, so I am a bit tired. Oh, I should mention that I got your packages (up to #14, I think?)—good idea with the Christmas cards! And tell Uncle John* (* in left margin of page: And Aunt Susan!) I said a big thanks for the maps! They came just in time—I was talking about America with grade 4/5/6 and grade 7/8. I was talking about earthquakes and plates—good timing! There is a letter addressed to Victoria (in D.C. …) that's for her and Uncle John and all those in said party. Anywho, I am teaching English songs right now. This has been a great addition to class. 4/5/6 = "You are my Sunshine," 7/8 = Blackbird" Compound words, figurative language, rhyme—man, it's great! And next semester … I got the Macdaddy of all songs planned. I am excited! I don't feel like sullying this good mood with talks of how bad the education is here. All I'll say is this—for the last 7-10 years for the kids in 7/8 grade, what have they been doing? But, so as not to sully (for sullying makes one sullen), I can only say this: we are pulling a Jeffersons. We may even get a piece of the pie. Other than school, I've managed to go on many adventures (crab hunts, making copra, night spear fishing this Saturday (!), trips to other islands), I've completed some books (War and Peace, Confederacy of Dunces, Pandora's Seed, Guns, Germs, and Steel, Modern Poetry (volume I), and I'm always reading new ones (currently: We, the Navigators, The Vintage Bradbury, and On the Road (the original manuscript!)). As far as basketball court news … I am waiting for information on a grant from a Marshallese group, CMI (the college) may want to donate, and a store called EZ-Price seemed interested to help. However, it has been difficult to maintain contact. I am going to keep trying, but I may not be in Majuro again until after Christmas. If people have donated to yall, that's fantastic! If anything, should a court fall through, it may be better to do repairs on the school, maybe a paint job—who knows. You'll know come December/January. I'm glad to hear people are doing well and that you had a "Halloweeny" night! Hopefully, my letter to André gets to her and Big J soon. Tell Steven to keep up the pace. I know he's kicking … booty. I also hope all is well on Monte Carlo. I miss all yall siblin's, but I know I have nothing to worry about. Ma and Pa, how's the house coming along (or is it finished?)? Any plans to move to Europe or anything? Maybe go visit André and John? I send all my love in laurels and bayleaves to all of you. I* also send my Happy Thanksgivings, Happy Hannukas, Happy Kwanzaas, Merry Christmases, happy birthdays, and Happy days in general. Eat turkey! I hope "Christmas ilo majel" works out! I want pumpkin pie … or pecan pie … haha! (* fading ink for the remainder of this letter:) Ok, this is far enough. My hand hurts. I love you, and I'll talk to yall soon!

Love,

James

(P.S. I'm fading away … .)

***

11/2/10

Dearest Jess, the honorable Luis, and my despicable chums,

Well, it's November. And I am still writing to yall cats. So you know yall good in my book. First, I'll start with a Jess and Luis aimed letter: guys, yall missin' out. The Pacific is where it's at. How is home? I crave a Jess cheese puff (and now I'm thinking of the League Party—what will yall do?). Jess, I turned on my cell the other day to see if it still worked; I looked at my texts and I remember I hate you. The text (…) when I was in Hawaii … unforgiveable. Lu, you treatin' your dame with love? How's bidness? Still in N.O. with the sis and bro? I had a dream the other night about that Carl, Jr.'s in Oregon we walked to, and the rodeo burgers we ate. But I also drank a coconut in that dream … so not exactly as I remembered. I'd kill for some black beans and tortillas—corn ones. I have flour in everything I eat here. Ai ramen—make ramen, add flour to thicken, and add salt to the mixture. I will have diabetes. Haha. I hope yall have gotten the previous updates. Ask Brett, Chris, and Doug if not—yall need to coordinate, dawgs. Oh, how my Saints; and how's LSU? And how is school, Jess? Do you graduate in May, '11? Ok, now for generalities and such. The Marshallese poop in the ocean. It's logical. But some poop on the beach. This makes it difficult walking at times, especially walking at night. Kubwe (koo-bwe) is poop. I've taught the kids "feces" as proper terminology. When we walk to school, I constantly hear "James! (or Shay-mez) Look out the feces!" maybe 18 times every morning. (…) I've been uber busy as of late. I'm trying to get a TEFL certification so I can be a licensed teacher (for foreign language teaching). It's a lot of work—sending lesson plans, answering "reflections," and tomorrow I need to film my class ad assess myself. Then I film again and reassess. All of this is coupled with things like maybe not having electricity on some nights, days of copious rain, having 11 little siblings, and making potato salad for my sister's kemem (the one year old birthday party). Days are moving so fast here—I mean November? So crazy. I hope everyone is well and happy. At least by December I may be able to do some more direct communication. But for now, I'm pretty content being pretty removed. I'm focused on what I have to do, yet I get to be all contemplative.

I'll quit here today. Need to think it out. Homies.

***

11/17/10

I've thought out. I felt like my life was like a Wu-Tang song. Quite deep. The kemem was postponed again, but it may be tonight. In defense of the Marshallese, it has been raining a lot. Like a lot. My feet get wet when I sleep. Womp, womp … Doug, I got your Halloween package yesterday! Late, sure, for fantastic nonetheless! And Sunday comics! An Alongia tradition, no doubt. Hope Finn (and you and the Missus) are healthy. And crazy about the Chimes. Poor Fred (…). How is the haunt now? Do yall still visit from time to time? I've been learning some Marshallese songs as of late. And a lot. So it's been cool because it's essentially my adult hangout time. We drink tea, coffee, or kool-aid, whip out guitars and ukuleles, and sing and talk. Otherwise, I'm with the kids. Which, admittedly, can be a little taxing. The along time thing is an impossibility. The other night when leaving the Church, I was surprised to be walking on the beach alone. I looked at the sky and flipped on the iPod. Literally, I just sighed and said out loud "finally!" Maybe 30 seconds later, two of my brothers popped out of the dark. They were waiting for me so I wouldn't have to walk home alone (demons …). It's sweet, and culturally important, and blah, blah blah. But I just wanted to stargaze in my own world for just 10 minutes … Haha, whatcha gonna do? Now, I'm going to play UNO with my other brother. Until I write again!

***

11/14/10

It's 6:30 in the morning. This is usually a good time to write. Only a few people are up, and light is coming in through my window. I'm sure I've mentioned it before, but this week I've had some out-there dreams. USA & RMI are combining. I've got this Spanglish thing goin' on, but with Marshallese. They are speaking English; I'm speaking Marshallese. I've been disillusioned in this place (about the educational system), which I should have been a while ago. So much work, as far as academics, needs to be done. I'll have to tell you in person if you care. I'm ready for Christmas so I can get a small break and reassess my teaching style and what I can actually do. But, one thing I am sure of: the kids are doing better, even if it may be .00001% better. Anywho, the kemem was a hoot! It got moved to Sunday last week, but it was cool to be part of the whole thing; helping in whatever way I can. And, since I had vinegar in my bag, we made badakdok (sp?), which is boiled pig's blood. Tasty stuff! I got to sing songs with my band, The Beatles, which is made of my baba Tarturin, Mon Kuk, Tondin, and me (and sometimes Jonwi). Singing, eating, one year olds—party central up in here. I went on jambos (trips) and made copra yesterday. Today, I will refuse a day of rest; going to school to do some work, make posters, and try to come up with teaching ideas. For now, as it has started raining, I will put my head back down.

I hope all is marvelous!

***

11/18/10

This will be the stopping point of the letter. I want to get it out to yall by at least mid-December. I think my baba is going to Majuro tomorrow, so it may leave by next week. Today was "just another day" in Bikarej. I taught all of the English classes today. So I'm a lil bushed. Grade K-1 and 2-3: letter recognition, "basic" phonics (I'm tying in Mr. Brown Can Moo to sounds of letters). Grade 4/5/6 … we're reading a piece of text from a book (…).

The Ministry of Education wants us to use this textbook that was made for immigrant children who move to the U.S. It's unreliable and stupid. And the stories include awesome lines, like from "Emily and Alice"—"Alice, you have a rainy head!" Shoot me.

But we're also working on grammar (which I'm teaching both English and Marshallese grammar ... yay!). Grade 7/8—Awesome textbook that has texts about earthquakes and Troy and all sorts of cool stuff. Again, also working on grammar (using 2nd grade level books—they never heard of this stuff—what have they been doing in school for the last 8-10 years?). We are practicing English songs—next semester should be fantastic! I have the best song we will learn. Can't tell you, but I'll record it! Too excited! I am now the owner of a pig (which I may have mentioned before?) and have two kittens. Going night spear fishing on Saturday, reading, writing, and of course: Thinking! All the time! I have been, strangely enough, getting little pieces of alone time. I go get lost in the jungle or go to the lagoon and contemplate all the while (or sometimes just exist in those places). When the light is just right, and I am in the jungle ... The greens just explode, there are these gorgeous red flowers that juxtapose the greens and browns. And these beautiful white birds will often roost in the trees, ever watchful of their jungle. Or ... sometimes. They are tasty, too. I feel healthy. I feel happy. I have no worries (or, at least, none of the kind of worries I had back home). Let me know how yall are! I'm just petting a kitten. So ... yeah. Anything wild going on? Or boring? Or in between? How are the families? And friends? Tell Hayes that there is a doctor here (or really, never here) that looks like him if he was Marshallese. Brett & Ali, I got the picture! Marley's in school? That ... is so weird. I refuse to get old! Tell Marley I love her! And Doug and Rachel, yalls packages have been a big hit (and so have your boxes—hey o!). And the Sunday comics are a great "Alongia touch." Tell Finn I love him, too! Luis and Jess—there better be no babies right now. But if there is, I love him/her, too. Luis, there is a guy who has heard of you here. He said you came in search of (...). Bad form, Luis. Go to New Guinea (Papua, that is). Chris and Sarah—no babies. But, again, tell him/her I love 'em too if there is one. Send my love to all. And my Merry Christmases! No League Party ... or at least not with me. Do it right! Keep it (...) CLASSY! I'll talk to yall soon. Letter ends now.

Laurels and Bayleaves,

James

***

11/3/10

Dear Mom and Dad,

I am sending this package in the hopes it makes it to you for Christmas! So Merry, Merry Christmas! I think this package, in theory, will leave Monday next week. I hope it makes it to you in time. Enclosed are various amimono. One of the small ones takes between 1.5 and 2 hours. They sell them for between $1 and $2 (sometimes 50 cents). I want you to distribute them as my gifts as follows: big ones for 1) mom and dad 2) Grandmother 3) Grandpa & Aunt D, 4)5)6)7) McCrary, Cardinale, Steven, Simone. There are 15 smaller ones, and one small star. Can you be sure to give one to Mrs. Carolyn and (if you are willing) one to Dr. Chicoine in the ANTH department, giving them the message thank you and Merry Christmas! (* Added in the left margin on the first page of this letter: Oh! And the Shackletons) The women who make these amimono are very skilled, so really *appreciate the time put in! Some of these were gifts to me, some I bought, and a few are from my mama and baba. They are very excited to send these to yall (and they made a very big deal of it—they really go 100% with this stuff). (* Added to the margin at the top of the 2nd page of the note: I had to reopen the package—one of my sisters gave you 2 more—a necklace and a bracelet!) I know Christmas will be a good time in the Marshalls. I am teaching two songs in English to the Church youth group: "Joy to the World" and "I Saw Three Ships." They will be performing on Arno (the island) on Christmas—and I'll be on stage with them! I hope the amimono help a little bit with the Christmas theme this year! I love yall, miss yall, and know you'll have a yule-tide blast. I love you, mom and dad! Thanks for all the support, the mail, and the love—and not just now, but throughout my life! Yall good people. Send my love to all!*

Yokwe and Love,

James

[The 3rd page of this note is James' hand-written note with the Marshallese words for "Joy to the World!"]

***

Al in Christmas (Christmas Song)
"Lañloñ ñan lal"

Lañloñ ñan lal
Christ ej itok.
Armej ren bok aer kiñ!
Na jikin in
ilo buruwo,
Ri-lan in lal ren al,
Ri-lan in lal ren al,
Ri-lan im lal im lal ren al!

You may know this song as
"Joy to the World!"

CPSIA information can be obtained
at www.ICGtesting.com
Printed in the USA
BVHW050546021120
592244BV00001BA/1